MATHEMATICS in GAMES, SPORTS, and GAMBLING

– The Games People Play

MATHEMATICS in GAMES, SPORTS, and GAMBLING

– The Games People Play

Ronald J. Gould

Emory University
Atlanta, Georgia, U.S.A.

CRC Press
Taylor & Francis Group
Boca Raton London New York

CRC Press is an imprint of the
Taylor & Francis Group an **informa** business

A CHAPMAN & HALL BOOK

Chapman & Hall/CRC
Taylor & Francis Group
6000 Broken Sound Parkway NW, Suite 300
Boca Raton, FL 33487-2742

© 2010 by Taylor and Francis Group, LLC
Chapman & Hall/CRC is an imprint of Taylor & Francis Group, an Informa business

No claim to original U.S. Government works

Printed in the United States of America on acid-free paper
10 9 8 7 6 5 4 3

International Standard Book Number: 978-1-4398-0163-5 (Hardback)

Visit the Taylor & Francis Web site at
http://www.taylorandfrancis.com

and the CRC Press Web site at
http://www.crcpress.com

Dedication

To Lyn,

For putting up with all of this writing and more.

Contents

List of Figures   xi

List of Tables   xv

Preface   xvii

Author Biography   xix

1   Basic Probability   1
   1.1   Introduction   1
   1.2   Of Dice and Men   2
   1.3   Probability   9
   1.4   The Laws Which Govern Us   12
     1.4.1   Dependent Events   15
   1.5   Poker Hands Versus Batting Orders   23
   1.6   Let's Play for Money!   28
   1.7   Is That Fair?   32
   1.8   The Odds Are against Us   39
   1.9   Things Vary   41
   1.10   Conditional Expectation   45

2   The Game's Afoot   49
   2.1   Applications to Games   49
   2.2   Counting and Probability in Poker Hands   49
   2.3   Roulette   54
   2.4   Craps   58
     2.4.1   Street Craps   60
     2.4.2   Casino Craps   61
     2.4.3   Other Bets   62
   2.5   Let's Make A Deal — The Monty Hall Problem   66
   2.6   Carnival Games   69
   2.7   Other Casino Games   72
     2.7.1   Caribbean Stud Poker   72

	2.7.2	Keno	74
	2.7.3	Blackjack	76
2.8	Backgammon		79
	2.8.1	Hitting Blots	81
	2.8.2	Off the Bar	81
	2.8.3	Bearing Off	83
	2.8.4	Doubling	86

3 Repeated Play   **93**

3.1	Introduction	93
3.2	Binomial Coefficients	94
3.3	The Binomial Distribution	103
3.4	The Poisson Distribution	113
3.5	Streaks — Are They Real?	120
3.6	Betting Strategies	124
3.7	The Gambler's Ruin	129

4 Card Tricks and More   **135**

4.1	Introduction		135
4.2	The Five-Card Trick		135
	4.2.1	Adding a Joker to the Deck	139
	4.2.2	More Variations of the Trick	140
4.3	The Two-Deck Matching Game		147
4.4	More Tricks		151
	4.4.1	Friends Find Each Other	152
	4.4.2	The Small Arithmetic Trick	153
	4.4.3	The Nine-Card Trick	154
4.5	The Paintball Wars		155

5 Dealing with Data   **159**

5.1	Introduction		159
5.2	Batting Averages and Simpson's Paradox		160
5.3	NFL Passer Ratings		166
5.4	Viewing Data — Simple Graphs		169
	5.4.1	Time Plots and Regression Lines	175
	5.4.2	When To Find the Regression Line	181
5.5	Confidence in Our Estimates		186
5.6	Measuring Differences in Performance		190
	5.6.1	Coefficient of Variation	194
	5.6.2	Relative Performance	196

6   Testing and Relationships   **201**

    6.1   Introduction 201

    6.2   Suzuki Versus Pujols 201

    6.3   I'll Decide If I Believe That 204

        6.3.1   Errors 207

        6.3.2   Summary of Hypothesis Testing 208

        6.3.3   One-Sided Tests 208

        6.3.4   Small Sample Sizes 209

    6.4   Are the Old Adages True? 212

        6.4.1   Home Field Advantage 212

        6.4.2   Lefty Versus Righty 215

    6.5   How Good Are Certain Measurements? 219

        6.5.1   Batting Average and Runs Scored 222

    6.6   Arguing over Outstanding Performances 224

    6.7   A Last Look at Comparisons 227

        6.7.1   Small Sample Comparisons 227

7   Games and Puzzles   **233**

    7.1   Introduction 233

    7.2   Number Arrays 233

        7.2.1   Magic Squares 233

        7.2.2   Variations on Magic Squares 239

        7.2.3   Sudoku 242

    7.3   The Tower of Hanoi 245

        7.3.1   Finding Solutions 248

        7.3.2   Bicolored Tower of Hanoi 251

        7.3.3   The Derangement Tower of Hanoi 253

    7.4   Instant Insanity 255

    7.5   Lights Out 262

    7.6   Peg Games 266

        7.6.1   English Board 267

        7.6.2   Triangular Peg Solitaire 272

8   Combinatorial Games   **277**

    8.1   Introduction to Combinatorial Games 277

    8.2   Subtraction Games 278

    8.3   Nim 280

        8.3.1   Poker Nim 286

        8.3.2   Moore's Nim 287

        8.3.3   Other Games 289

x

8.4 Games as Digraphs 291
    8.4.1 Sums of Games 292
    8.4.2 The Sprague-Grundy Function 293
    8.4.3 More about Impartial Games 296
8.5 Blue-Red Hackenbush 298
8.6 Green Hackenbush 306
    8.6.1 Pruning Green Hackenbush Trees 306
8.7 Games as Numbers 310
8.8 More about Nimbers 315

9 Appendix   **319**
9.1 Review of Elementary Set Theory 319
9.2 Standard Normal Distribution Table 326
9.3 Student's *t*-Distribution 327
9.4 Solutions to Problems 328
9.5 Selected Even Exercises 331

References   **347**

Index   **351**

List of Figures

1.1   Girolamo Cardano. (Image from Wikipedia.)   2
1.2   Galileo. (Image from Wikipedia.)   3
1.3   Blaise Pascal. (Image from Wikipedia.)   4
1.4   Pierre de Fermat. (Image from Wikipedia.)   5
1.5   The choice tree for the Multiplication Principle.   6
1.6   The tree diagram for Example 1.4.8.   20
1.7   The dominance relations for Penny Ante.   36
1.8   The general tree diagram.   37
1.9   The right subtree in more detail.   38

2.1   American Roulette wheel.   55
2.2   Roulette table.   57
2.3   A typical craps table.   59
2.4   Backgammon board in opening position.   80
2.5   Example 2.8.3 end game position.   84
2.6   Example 2.8.4 end game position.   84
2.7   Example 2.8.5 end game position.   85
2.8   An end game position, do you double?   87
2.9   Another end game position, do you double?   88

3.1   Probabilities of i tails in eight flips.   104
3.2   Probability distribution for two flips.   105
3.3   Probability distribution for four flips.   105
3.4   Approximating bell curve for probability distribution. .   106
3.5   Area under bell curve when $Z = 2.12$.   108
3.6   We seek area to the right of $x = Z$.   108
3.7   Example of a Z score that is negative.   109
3.8   Symmetry allows us to obtain wanted value.   110
3.9   Area for Example 3.3.6.   111

4.1   The five-card trick.   136
4.2   The card rank circle (standard deck).   138
4.3   The five-card trick completed.   139

4.4   A drawing for the graph of Example 4.2.4. 143
4.5   The bipartite graph obtained. 145
4.6   You determine the face-down card. 147
4.7   The strategy tree diagram. 156
4.8   The left subtree. 157

5.1   Ripken time plot of home runs. 176
5.2   Time plot with well-placed straight line. 176
5.3   Ripken time plot for restricted period. 177
5.4   Gretsky goal-scoring plot. 180
5.5   Data set that is not approximately linear. 182

6.1   A time plot of batting averages for Pujols and Suzuki. . 203
6.2   Home runs versus runs scored, AL 2008. 221
6.3   Home runs versus runs scored, all of MLB 2008. 223
6.4   Team batting averages versus runs scored, MLB 2008. . 223

7.1   The Tower of Hanoi puzzle. 246
7.2   The n-cubes Q_n, $n = 1, 2, 3$. 248
7.3   Another graph model. 250
7.4   The 2-disk graph model. 251
7.5   The 3-disk graph model. 252
7.6   A longest solution for 3 disks. 253
7.7   A 2-disk bicolored Tower of Hanoi. 254
7.8   A 2-disk derangement Tower of Hanoi. 254
7.9   The standard representation for a cube. 256
7.10 The four cubes. 257
7.11 The corresponding graphs. 258
7.12 The superimposed graph S. 259
7.13 The two subgraphs S_1, S_2. 259
7.14 The cubes of Exercise 7.4.1. 260
7.15 The cubes of Exercise 7.4.2. 261
7.16 The cubes of Exercise 7.4.3. 261
7.17 A 2×2 Lights Out grid graph. 263
7.18 Mini Lights Out model. 265
7.19 English board (A) and European board (B). 267
7.20 Labeling the holes. 269
7.21 Eleven holes labeled y and the five that are possible solutions. 270
7.22 A 6-purge (i), 2-package (ii) and 4-package (iii). 272

7.23 A 3-purge (iv) and an L-purge (v). 272
7.24 A package solution to the central puzzle. 273
7.25 A labeled triangular board. 273
7.26 Labeling of the triangular board. 274
7.27 Solution to the English board with $(-1, 1)$ open. 276
7.28 Solution to the English board with $(0, 1)$ open. 276

8.1   An example of Northcott's game. 289
8.2   The Silver Dollar game. 290
8.3   The digraph of the subtraction game. 292
8.4   The digraph when $S = \{1, 3\}$. 293
8.5   Two Sprague-Grundy graphs. 297
8.6   A blue-red Hackenbush game. 299
8.7   A blue-red Hackenbush example. 299
8.8   A test of the one edge advantage hypothesis. 300
8.9   A test for the value $1/2$. 300
8.10 An advantage of 1.5. 301
8.11 The Hackenbush string for $2\frac{3}{4}$. 303
8.12 Hackenbush string for $-\frac{1}{4}$. 304
8.13 Hackenbush string for $-1\frac{9}{16}$. 304
8.14 Four examples of Hackenbush strings. 305
8.15 Generalized Hackenbush game. 305
8.16 Green Hackenbush trees. 306
8.17 Step 1 in finding the equivalent Nim pile. 308
8.18 Step 2 in finding the equivalent Nim pile. 308
8.19 Five Hackenbush trees. 309
8.20 A Hackenbush game. 309
8.21 Another Hackenbush game. 310
8.22 (a) A fuzzy game and (b) the sum of two fuzzy games. . 314

9.1   Venn diagram for $A \cup B$. 322
9.2   Venn diagram for $B \cap \overline{A}$. 323
9.3   Venn diagram for $\overline{A} \cap \overline{B}$. 324
9.3   Standard Normal Distribution: values correspond to area
      shown in figure. 326
9.3   Student's t distribution (from [39]). 327
9.4   Labeling for Problem 7.6.1. 329
9.5   The choice tree for Exercise 1.2.2. 331
9.6   The cubes of the solution for Exercise 7.4.2. 342
9.7   The cubes of the solution for Exercise 7.4.4. 342

9.8   Venn diagrams for Exercise A.1.2.   346

List of Tables

1.1   Sample Space of Pairs When Rolling Two Dice   7
1.2   The Sample Space of Sums When Rolling a Pair of Dice   8
1.3   Payment Table for the Coin Flip Game   29
1.4   Probabilities of the Outcomes in S   29
1.5   The Possible Outcomes with Further Play   32
1.6   The Patterns, Some Responses and Probabilities   35

2.1   Counting Five-Card Poker Hands   51
2.2   Counting Three-Card Poker Hands   53
2.3   True and House Odds in American Roulette   56
2.4   Partial Table for Events and Probabilities for Craps . .   60
2.5   Odds and Casino Edge for Some Other Craps Bets . . .   64
2.6   Pay Table for Raises in Caribbean Stud Poker   73
2.7   Some Keno Bets and Payoffs   75
2.8   Probabilities of Hitting a Single Blot with One Piece . .   82
2.9   Probabilities of Entering from the Bar   82

3.1   Probabilities for Exactly r Tails in Eight Flips of a Fair Coin   96
3.2   Pascal's Triangle   100
3.3   Binomial and Poisson Values for $P(k)$, $0 \leq k \leq 8$ 116

4.1   A Matching in the Bipartite Graph   146

5.1   Comparison of Ronnie Belliard and Casey Blake   162
5.2   Comparison of Derek Jeter and David Justice: 1995– 1997   163
5.3   Breakdown of Hitting Statistics for Two Players   164
5.4   Home Run Totals for Both Leagues in 2007   170
5.5   Stem Plot of Major League HR Totals   171
5.6   A Comparison Stem Plot of the Home Run Data   172
5.7   Team Batting Averages for 2007   173
5.8   Stem Plot for Team Batting Averages in 2007   173

5.9   Two-Decade Comparison of American League Leading
       Averages 174
5.10 Cal Ripken's Home Run (HR) Totals by Year 175
5.11 Ripken Home Run Totals, 1986–1989 177
5.12 Comparison of Points Scored by Oscar Robertson and
       Jerry West 184
5.13 Top National League Averages for the 1920s and 1980s . 184
5.14 Typical Z Values 187
5.15 Hits for American League Players with 300 At Bats in
       2007 192

6.1   Pujols Versus Suzuki, Batting Average Stemplot 202
6.2   2008 Home and Away Records for AL 213
6.3   2007 Home and Away Records for AL 214
6.4   Percentage Differences for 2008 215
6.5   2007–2008 Differences in Averages 217
6.6   2008 Team Offensive Statistics 220
6.7   Stemplot of Hits per Season, Pujols Versus Suzuki . . . 228

7.1   Loh Shu Magic Square 234
7.2   A Magic Square 235
7.3   The Complementary Square 236
7.4   The Number of Small Magic Squares 237
7.5   An Unknown 3 × 3 Magic Square 237
7.6   System of 8 Equations in 9 Unknowns 237
7.7   A Weak Magic Square with Modified Entries 241
7.8   A Pair of Orthogonal Latin Squares 241
7.9   A Partial Sudoku Array 243
7.10 A 6 × 6 Sudoku Puzzle 244
7.11 Lights Out Game 263
7.12 Typical Corner, Edge and Center Move Arrays 264
7.13 Addition Table for the Special Group of 4 Elements . . 268

8.1   Four Possible Game Types 311
8.2   Four Possible Game Types Re-Examined 313

Preface

This book has grown out of a freshman seminar that I have taught regularly for over ten years. As a freshman seminar, it requires only high school algebra as a prerequisite. We will encounter several somewhat deeper mathematical topics, but will handle them at a level commensurate with the prerequisites.

This book is intended to draw the reader's interest to applications of mathematics, especially to topics in which many people already have some interest, namely games, sports and gambling. My hope in creating this course was that students would be more excited and interested in the applications and, because of that, they would become more interested in the mathematical theory. To help promote this, I have tried to keep all the examples, questions, and problems within the realm of games, sports and gambling. I have also tried to obtain real life data for as many of the problems as possible. The Web is an easy source of such data. I have taken a few minor liberties such as assuming a batting average was a probability and that sequences of at bats were independent events. This was done to expand the example set.

The text is built around numerous examples, problems and questions. Examples have been solved in detail to allow the reader to gain a firm understanding of the material. Problems are intended for the reader to solve. I use them as in-class group problems or occasional homework problems. Questions are of a broader nature and some are quickly answered, others more slowly answered and some are used simply to move us towards better questions. I do this as a way of showing students that by asking questions we can often guide ourselves to the correct question and then to a solution for that question.

Having had so many chances to "play" with this material, I have experimented with a number of different versions of the course. Often I have also tried to let the interests of that particular group sway the development of course material. Some groups are more interested in sports than games and some groups are the exact opposite. Eventually, this caused me to have far more material available than I could expect to teach in any one course. That is evident in the book. Hopefully this will allow you to shape the material to your interests or those of your students.

This is really a book about elementary discrete probability and statistics, along with some discrete mathematics. It is possible to keep the course centered around probability and statistics (Chapters 1, 2, 3, 5, 6) without ever bringing games, puzzles and card tricks (Chapters 4, 7, 8) in at all. Or one could have a basic discrete probability and discrete math course, using Chapters 1, 2, 3, 7 and 8. Chapters 2 and 6 include a number of applications to material from Chapters 1 and 5, respectively.

Because there is so much flexibility in what can be covered, I have indicated some sections that can easily be omitted. These include Sections 1.10, 2.6, 2.7, 3.6, 3.7, 4.4, and 4.5. Chapter 5 introduces elementary statistics and Sections 5.2 and 5.3 can be omitted. However, these are two sections students enjoy. In Chapter 6, Sections 6.3 and 6.7 do fundamental new statistical topics while the other sections concentrate on applications of topics already learned. In Chapter 7 every section stands alone, so you may do them in any order or not at all. However, Chapter 8 builds section by section and so should be covered that way.

I have also included a brief review of set theory in Appendix A.1. This sets the tone for sample spaces and events. It also ensures that everyone has seen and is familiar with the standard notation in use in the text. I have found a day or two on this review always helps. I did not worry about making all the examples in the Appendix come from games, sports and gambling.

The course uses many different games as examples. I have tried to include enough of a description of each game to make questions understandable, without creating a big book of rules. Hopefully I have shown enough examples to peak your interest.

I would also like to thank a number of people who have been helpful in the production of this book. These include Bob Stern, for believing in the project in the first place, and Jennifer Ahringer and Michele Dimont, for their kind editing assistance, and Ken Keating and Shashi Kumar for their technical help. Also, special thanks to my wife Madelyn Gould, first, for proofreading sections of the book, and second, for having the patience to put up with my writing another book. Thanks to Jim Albert for many helpful suggestions. I would also like to thank Kinnari Amin for her careful reading and many useful suggestions on the text.

Author Biography

Ronald J. Gould received a B.S. in mathematics from the State University of New York at Fredonia in 1972, an M.S. in computer science in 1978, and a Ph.D. in mathematics in 1979 from Western Michigan University. He joined the faculty of Emory University in 1979.

Dr. Gould specializes in graph theory with general interests in discrete mathematics and algorithms. He has written over 135 research papers and 1 book in this area. Dr. Gould serves on the editorial boards of several journals in the area of discrete mathematics. Over the years he has directed over 2 dozen master's theses and more than 20 Ph.D. dissertations.

Dr. Gould has received a number of honors including teaching awards from Western Michigan University (1976) and Emory University (1999), as well as the Mathematical Association of America's Southeastern Section Distinguished Teaching Award in 2008. He has also received alumni awards from both Fredonia and Western Michigan University. He was awarded the Goodrich C. White Chair from Emory University in 2001.

Chapter 1

Basic Probability

1.1 Introduction

This text is intended to demonstrate some of the mathematical principles underlying many of the games we watch or play. These include, but are not exclusive to many of the presently very popular gambling games. Here we will build the laws of probability and show some of the fundamental consequences of this theory. Examples from sports will be used to build some of the concepts of statistics. We hope to use these examples to demonstrate what statistics is meant to study, why such ideas are useful and how these insights can really help us. Finally, we shall consider a number of games and other diversions such as card tricks that are mathematically based, or can be studied from a mathematical perspective. These other games and diversions will be used to demonstrate additional mathematical principles. Examples will be kept as close to real-life situations as possible, but occasionally small liberties will be taken in order to simplify the study.

The hope is that these real and unusual examples will bring the reader a broader appreciation of mathematics through the principles underlying so much of the world around us. Hopefully, it will also be fun to consider real games people play.

Exercises will be provided for each section, so that this book might serve as a text for a class and generally to help the reader understand the various concepts.

1.2 Of Dice and Men

There is a rich and interesting history to the games people play. Gambling can be traced back centuries. An early form of dice has been commonly found in Assyrian and Simerian archeological sites. Pieces used for "board" type games have been found in Babylonian and early Egyptian sites (see [8]). The Egyptians, Greeks and Romans all believed gambling had divine origins [36].

But for our purposes, the mathematical story really begins much later. Girolamo Cardano (1501–1576) was probably the first to write about the mathematics behind dice outcomes. He wrote "... before agreeing to stakes one must consider the total number of outcomes and compare the number of casts that would produce a favorable outcome to those that are unfavorable. Only in this proportion can mutual wagers be laid so that one can contend on equal terms" (see [36]). This was revolutionary for his time and remains a working definition for probability. Unfortunately, his writings on the subject were not published until nearly a century after his death in 1576 (see [36]). By this time, others had also written about the mathematics of dice games. Between 1613 and 1623, Galileo considered some questions involving

FIGURE 1.1: Girolamo Cardano. (Image from Wikipedia.)

mathematics' role in gambling and especially dice games. Writing in *Concerning an Investigation of Dice (Sopra le Scoperte dei Dadi)*, he began considering the mathematics of dice games. In particular, he noted the number of different sums that were possible when two or three dice were used and further, how many ways these various sums could occur (see [36]). These were fundamental steps in the proportion process noted by Cardano.

FIGURE 1.2: Galileo. (Image from Wikipedia.)

Although not the first to be caught up in gambling, Antoine Gombaud, The Chevalier de Méré (1607–1684) was a writer and gambler with a great deal of gaming experience. He used the name Chevalier de Méré in his writings and his friends eventually began calling him by that name. His gambling experience raised questions and he wondered about possible explanations (mathematical in nature) to these questions. Accounts vary somewhat as to the questions he actually asked about gambling. One common game at which he had considerable success was an even money bet on rolling at least one six in four rolls of a die. Some accounts (see [14], [32]) say because of his success at this game, he reasoned that betting on one or more double 6s in 24 rolls of two dice should also be a profitable game. However, he only realized through playing experience (and losing money), that he was not doing well with this game.

Other accounts (see [36]) say de Méré was questioning the *Problem of Points* (see Section 1.7). This is really a question of fair division in an interrupted game. The question is how to divide the prize fairly based on the present scores, when a game is only partially completed. The Problem of Points was first published in 1494 by Fra Luca Paccioli, a Franciscan priest, mathematician, teacher and friend of Leonardo da Vinci [36] . (Fra Luca Paccioli is now known as the father of modern accounting.) Thus, the problem had long bedeviled some of the best scholars.

FIGURE 1.3:   Blaise Pascal. (Image from Wikipedia.)

What question was really bothering de Méré is not completely clear from the literature, but what is clear is that in 1654 The Chevalier de Méré asked his friend, the well-known mathematician Blaise Pascal to help him understand one or both of these questions. Pascal clearly took an interest in the Problem of Points, as well as the other dice games mentioned above, and in a series of letters with another famed mathematician, Pierre de Fermat, de Méré's questions were eventually explained. In the process, the idea of probability was born, or at least finally formally stated in public. Pascal went on to develop his famed triangle and ideas of the binomial distribution (see Sections 3.2 and 3.3

for more information).

We are now ready to state our first major question.

Question 1.2.1. *Why was the first game profitable and why was The Chevalier de Méré losing money playing the second game?*

FIGURE 1.4: Pierre de Fermat. (Image from Wikipedia.)

We postpone a discussion of the Problem of Points until Section 1.7. To understand the real difference between the other two games de Méré questioned, we must put them in a proper mathematical setting. This setting we will call the *sample space*. The sample space is just the set of all possible outcomes of a certain experiment. Thus, let us suppose the first experiment is playing the first game, that is, rolling one die four times.

Each time one die is rolled, there are six possible outcomes, and assuming the die is fair (we will worry about unfair dice later), those six outcomes, which form the sample space, are equally likely to happen. We also note that rolling a die repeatedly creates *independent events*. That is, the die has no memory, so what happened on previous rolls has no effect on the next roll. When events are independent, there is a nice way to count how many ways outcomes might happen. This is called the *Multiplication Principle*.

Rule 1.2.1. The Multiplication Principle: *If task 1 can be performed in n ways and task 2 can be performed in m ways, then the number of ways of performing task 1 and then task 2 is the product nm.*

As a simple example we ask about the number of outcomes for flipping a coin two times. On any flip we can either get a head (H) or a tail (T). Because there are two possible outcomes on each of the two flips, the Multiplication Principle says there are four outcomes to this experiment. We can verify this by simply listing the possible outcomes. These are the *ordered pairs* (H, H) or (H, T) or (T, H) or (T, T).

We can picture the Multiplication Principle using a *choice tree* diagram (see Figure 1.5). Think of S as the starting point, before we perform either task. The points labeled $1, 2, \ldots, n$ correspond to the choices for task 1. Now, for each outcome i from task 1, there are m choices for task 2 labeled $(i, 1), (i, 2), \ldots, (i, m)$. Thus, there are $n \times m$ possible outcomes.

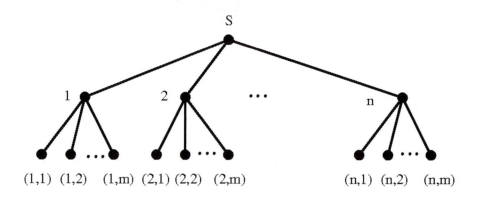

FIGURE 1.5:   The choice tree for the Multiplication Principle.

Thus, we can actually see the Multiplication Principle in action. This is an important rule and one we shall use often. In fact, the rule is not limited to two tasks, but can be done more generally.

Rule 1.2.2. The General Multiplication Principle: *For k in-*
dependent tasks, where task T_i can be done in n_i ways, the number of
ways to perform all k tasks is the product $n_1 n_2 \ldots n_k$.

Now we can explain the solution to de Méré's dice problems. In the
one-die game, using the Multiplication Principle, there are a total of
$6^4 = 1296$ possible outcomes. That is, four rolls of one die with six
outcomes per roll. But how many of these contain at least one 6? It
is easier to count those that have no 6s. If we are not allowed a 6,
then there are only five outcomes per roll and so there are $5^4 = 625$
outcomes with no 6. Thus, the outcomes with at least one 6 total
$1296 - 625 = 671$. Thus, the proportion (remember Cardano!) of
outcomes with at least one 6 is

$$\frac{671}{1296} = .5178 > \frac{1}{2}.$$

That is, there are more ways to obtain at least one 6 than there are
ways to obtain no 6s. Hence, betting on at least one 6 should mean
you win more often than you lose.

TABLE 1.1: Sample Space of Pairs When
Rolling Two Dice

	1	2	3	4	5	6
1	(1,1)	(1,2)	(1,3)	(1,4)	(1,5)	(1,6)
2	(2,1)	(2,2)	(2,3)	(2,4)	(2,5)	(2,6)
3	(3,1)	(3,2)	(3,3)	(3,4)	(3,5)	(3,6)
4	(4,1)	(4,2)	(4,3)	(4,4)	(4,5)	(4,6)
5	(5,1)	(5,2)	(5,3)	(5,4)	(5,5)	(5,6)
6	(6,1)	(6,2)	(6,3)	(6,4)	(6,5)	(6,6)

But what about the second game, with a pair of dice rolled 24 times?
When a pair of dice are rolled, the Multiplication Principle tells us
that the sample space consists of $6^2 = 36$ different outcomes (see Table

1.1). Of these outcomes, only one is double 6s. Thus, 35 out of the 36 outcomes of our task are bad for us. Considering this fraction then, the proportion of outcomes with no double 6s is

$$\left(\frac{35}{36}\right)^{24} = .5086 > \frac{1}{2}.$$

That is, more than half of the outcomes have no double 6 in 24 rolls of a pair of dice. Thus, these are losing outcomes in game 2 and we should expect to lose this game more often than win.

The Chevalier de Méré had his sad news. His reasoning had been flawed and the truth could be seen by a careful examination of the sample spaces for the two games and then determining the proportion of favorable or non-favorable outcomes in each space.

The reader should be aware that the form of the sample space is all up to you. It should be based on the experiment at hand and the information needed. We have seen one sample space for rolling a pair of dice, but the same general experiment could have other sample spaces. For example, if we roll a pair of dice and want to know the total sum rolled, then our sample space could be as shown in Table 1.2.

TABLE 1.2:   The Sample Space
of Sums When Rolling a Pair of Dice

die 1 → die 2 ↓	1	2	3	4	5	6
1	2	3	4	5	6	7
2	3	4	5	6	7	8
3	4	5	6	7	8	9
4	5	6	7	8	9	10
5	6	7	8	9	10	11
6	7	8	9	10	11	12

Exercises:

1.2.1 How should The Chevalier de Méré expect to do if his original one die game had only three rolls in order to roll a 6? What if he had five rolls in order to roll a 6?

1.2.2 What is the size of the sample space for a game where the outcomes are: You roll a die and then flip a coin? Create a choice tree to verify your answer.

1.2.3 How might you represent the outcomes for the game in the previous problem?

1.2.4 What is the sample space for the experiment of drawing one card from a standard deck of 52 cards? (Note: a standard deck will always have 52 cards composed of 4 suits, each with 13 cards, ace, two, ... , ten, jack, queen, and king.) .

1.2.5 What is the sample space for drawing one card if our deck consists only of all the hearts and all the aces from a standard deck?

1.2.6 How large is the sample space for the experiment of rolling one die, then drawing one card from a standard deck?

1.2.7 Create a choice tree to show the possible outcomes for flipping a coin three times. Verify the number of outcomes by using the Multiplication Principle.

1.2.8 Create a choice tree to show the possible outcomes from the experiment of flipping a coin, then rolling one die, then flipping a second coin. Use the Multiplication Principle to verify the number of outcomes.

1.3 Probability

Now that we have the idea of a sample space, we can define exactly what we mean by *probability*. This is done in the same set theoretic terms that we used for the idea of sample space. An *event* will simply be a subset of the sample space. So an event is just a collection of possible outcomes. For example, if our experiment is rolling one die, some possible events include $E_1 = \{6\}$ or $E_2 = \{2, 4, 6\}$ or $E_3 = \{1, 6\}$. So event E_1 is rolling a 6, while event E_2 is rolling an even number and event E_3 is rolling a 1 or a 6.

Given an event $E = \{a_1, a_2, \ldots, a_k\}$, the probability of event E, denoted $P(E)$, is defined to be:

$$P(E) = \sum_{i=1}^{i=k} P(a_i). \tag{1.1}$$

That is, the probability of an event is just the sum of the probabilities of the individual outcomes that comprise the event.

Suppose we would like to roll an even number when rolling one die (event E_2 from earlier). Then, as we are assuming an unbiased die, we can compute the probability of event E_2 by summing the individual probabilities of the outcomes that comprise E_2. Thus,

$$P(E_2) = P(\{2, 4, 6\}) = 1/6 + 1/6 + 1/6 = 1/2.$$

Similarly, $P(E_1) = 1/6$ and $P(E_3) = 2/6 = 1/3$. Thus, our sample space for rolling a pair of dice makes these probabilities easy to determine.

We can also use the sample spaces for rolling a pair of unbiased dice to see more examples of probability computations. From Table 1.1 we can see that P(rolling double sixes) $= 1/36$, as there is only one such outcome in the 36 possible outcomes. What is the probability of rolling a total of 7 with two dice? From Table 1.2 we can see that this is

$$P(\text{rolling a total of 7}) = 6/36 = 1/6.$$

As another example of this idea, again consider our example of flipping a coin two times. We know the sample space of this experiment consists of exactly four outcomes $(H, H), (H, T), (T, H), (T, T)$ and thus, if our coin is assumed to be unbiased, each of these outcomes is equally likely. This means that the probability of flipping two consecutive heads should be $\frac{1}{4}$. (Note: this equals the value obtained from the Multiplication Principle using probability 1/2 for each of the two rolls.)

What happens to our sample spaces if we now wish to roll three dice? If we are considering the individual values on the three dice, then from the Multiplication Principle we know this set has $6^3 = 216$ outcomes. Thus, we certainly do not wish to write them all out as we did for two dice in Table 1.2. But how can we think about this sample space? Probably the easiest way is to think of the three dice as being of different colors, say red, white and blue. Then we can record the outcomes of rolling the three dice based on their colors. If when we roll, the red die is 6, the white die is 4 and the blue die is 3, then we could write this outcome as $(6, 4, 3)$. That is, this *ordered triple* records that particular outcome perfectly. The set of all 216 of these ordered triples

$$(\text{red die}, \text{white die}, \text{blue die})$$

comprises our sample space. We have now rediscovered what Galileo had noticed 400 years ago!

So ordered pairs were used to record the outcomes from flipping a coin twice. Now ordered triples help us record the outcomes when we roll three dice. In general, an ordered k-tuple $(x_1, x_2, \ldots, x_k)$ can be used to show the results of $k \geq 2$ experiments.

Thus, we have seen the use of ordered k-tuples to be a practical way of recording outcomes of multi-events. Flipping a coin two times can be recorded as ordered pairs

$$(\text{first flip}, \text{second flip})$$

and, in general, rolling k dice can be recorded as ordered *k-tuples*. This form of modeling our sample space will continue to be useful throughout the text.

Exercises:

1.3.1 Create a table for the probabilities of each possible sum when rolling a pair of dice.

1.3.2 What is the probability of drawing a spade from a standard deck? What is the probability of drawing an ace from a standard deck?

1.3.3 What is the probability of drawing a king or a queen from a standard deck?

1.3.4 Flip a coin and draw a card from a standard deck, now determine the size of the sample space for this experiment. What is the probability of any single element in this sample space?

1.3.5 What is the probability of drawing a heart that is not an ace from a standard deck?

1.3.6 What is the probability of drawing a diamond that is not a jack, queen or king from a standard deck?

1.3.7 Suppose you play the following game: you roll one die and you flip one coin. Winning payments are as described in the table below. Determine the probabilities for each of the winning payments.

	1	2,3,4	5,6
H	3	0	2
T	2	1	1

1.3.8 What is the probability of first rolling a 6 with one die, then drawing a club from a standard deck?

1.3.9 You first flip a coin, then roll a die, then draw a card from a standard deck.

1. What is the probability of obtaining a head, a 3 and a club?
2. What is the probability of obtaining a tail, an even number, and an ace?
3. What is the probability of obtaining a tail, a number at least as large as 3, and a jack, queen or king?

1.4 The Laws Which Govern Us

In this section we will develop some of the fundamental laws of probability. The reader well versed in probability theory might skip this section. Those just learning probability will wish to go carefully through this section.

In the last section we took for granted several important rules. We state them here for completeness.

Rule 1.4.1. *The sum of the probabilities of all events in the sample space equals one.*

This is fundamental. A probability is a proportion of the time that an outcome should occur. We cannot then exceed one in summing these proportions. If we fail to reach one, then some outcome has been overlooked or some probability of an outcome(s) is incorrectly given.

Rule 1.4.2. *If event E has probability p, the probability event E does not happen is* $1 - p$.

This is a very useful rule and a consequence of the first rule. Sometimes it will be easier for us to compute the probability of the failure of the event E, which we denote as $\overline{E}$ (the standard set complement of the set E), rather than computing the probability of success for event E. Thus, restating this rule we have

$$P(E) = 1 - P(\overline{E}),$$

which makes finding $P(E)$ easy when finding $P(\overline{E})$ is also easy.

Rule 1.4.3. *When an experiment has n equally likely possible outcomes, then the probability of any one of them happening is* $\frac{1}{n}$.

When we have this situation, we say that we have a *uniform probability distribution*, or we often will just say, a *uniform distribution*. When we have a uniform distribution, then we merely need to count the possible outcomes that satisfy any event of interest in order to compute the probability of the event happening. Thus, methods of counting will become important techniques and we will spend a great deal of energy learning counting techniques. Of course, not all probability distributions are uniform.

We have already seen examples of uniform distributions. The probability distributions for rolling an unbiased die, or flipping a fair coin twice are examples of uniform distributions. We made all our initial assumptions based on this model.

Our next rule is a natural consequence of Rule 1.4.3 and stresses the need to be able to count the number of outcomes of certain experiments.

Rule 1.4.4. *In a uniform distribution with n equally likely outcomes, if some event E corresponds to the union of k of these outcomes, then* $P(E) = k/n$.

Example 1.4.1. *Suppose we toss a fair coin three times. What is the probability we get exactly two heads?*

Solution: Here we need to first determine the size of our sample space. In this case it will have $2^3 = 8$ equally likely possible outcomes. We view these outcomes as 3-tuples representing the outcomes of the three tosses. Thus, we must decide how many of these 3-tuples have exactly two heads. But this is possible by a simple inspection of the list of possible 3-tuples.

That list is:

$$(H,H,H) \quad (H,H,T) \quad (H,T,H) \quad (T,H,H)$$
$$(H,T,T) \quad (T,H,T) \quad (T,T,H) \quad (T,T,T).$$

Thus, we see that there are three such 3-tuples and hence

$$P(\text{exactly 2 heads in 3 tosses }) = \text{ number of such outcomes } \times \frac{1}{8} = \frac{3}{8}. \quad \square$$

Clearly, the inspection method will become a problem if we want to do 50 tosses! Thus we need better ways to count such things and in the next section we shall begin attacking this question.

Rule 1.4.5. *If E_1 and E_2 are disjoint (also called mutually exclusive events), then the probability of event E_1 or E_2 happening is the sum of their individual probabilities. That is*

$$P(E_1 \cup E_2) = P(E_1) + P(E_2).$$

To see that this rule makes sense just think of the "or" condition as used in set theory. We are content to have any outcome that occurs in event E_1 "or" any outcome from event E_2. As these events share nothing in common, this is the same as taking the union of the two events to form one larger event. This new larger event clearly has probability the sum of the probabilities of the two smaller events. That is, $P(E_1 \cup E_2) =$ the probability of E_1 happening plus the probability of E_2 happening.

Example 1.4.2. *What is the probability of drawing a 7 or drawing an 8 from a standard deck of 52 cards?*

Solution: Here we have two events, let E_1 be the event of drawing a 7 while E_2 the event of drawing an 8. These are exclusive events and thus,

$$P(E_1 \cup E_2) = P(E_1) + P(E_2) = 4/52 + 4/52 = 8/52 = 2/13. \quad \square$$

Example 1.4.3. *What is the probability of rolling a 2 or a 4 with one die?*

Solution: Again our two events are exclusive and so

$$P(2 \text{ or } 4) = \frac{1}{6} + \frac{1}{6} = \frac{1}{3}. \quad \square$$

Example 1.4.4. *What is the probability of drawing a spade or drawing a king when drawing one card from a standard deck?*

Solution: Our two events, drawing a spade or drawing a king are not exclusive! Thus, we must consider the situation more carefully and adjust. As there are 13 spades and one of them is a king, there are only three other kings not included among the spades. Now, these two new events are independent and our computation becomes

$$\frac{3}{52} + \frac{13}{52} = \frac{16}{52} = \frac{4}{13}. \quad \square$$

This example brings us to the next idea.

1.4.1 Dependent Events

As we just have seen, some events certainly do have an impact on other events. For example, before a baseball game we might feel the starting pitcher has a 2% chance of throwing a nine inning shutout. We might also feel that his team has a 70% chance of winning the game. However, the chance of the pitcher winning the game and also pitching a shutout is not 70%, but still more like 2%.

As another example, you might ask what is the probability the next card you draw is an ace, if two aces have already been dealt? Clearly, the information that two aces have already been dealt changes the probability you seek.

Events like these, namely events that affect the probability of later events, are called *dependent* events. More formally, event A is dependent upon event B if the probability of A is affected by the occurrence or nonoccurrence of B. Given dependent events E_1 and E_2, we may wish to consider the *conditional probability* of E_2 given E_1, which is the probability that if E_1 happens, E_2 will also happen.

Formalizing our notation we will use the following:

- $P(E_2 \mid E_1)$ = the conditional probability of E_2 happening given that E_1 has already occurred;

- $P(E_1 \cap E_2)$ = the probability of E_1 and E_2 happening.

Then the following rules hold.

Rule 1.4.6. *Let E_1 and E_2 be events. Then*

1. $P(E_1 \cap E_2) = P(E_1)P(E_2)$ *for independent events.*

2. $P(E_1 \cup E_2) = P(E_1) + P(E_2) - P(E_1 \cap E_2)$ *for all events.*

3. $P(E_1 \cap E_2) = P(E_1)P(E_2|E_1)$ *for dependent events.*

Since two independent events must both happen, Rule 1.4.6 (1) is really just the Multiplication Principle in action. Rule (2) is a natural adjustment when events are not exclusive. We must adjust for the outcomes contained in both events and hence the fact that such outcomes are being counted in each event. The solution to this double counting is to subtract the probability of the common outcomes $(E_1 \cap E_2)$. This rule reduces to Rule 1.4.5 when the events are mutually exclusive.

It is also not difficult to see why (3) holds. Simply think of $P(E_2 \mid E_1)$ as the proportion of those outcomes satisfying E_2 out of all those outcomes already satisfying E_1. Rewriting (3) so that we solve for $P(E_2 \mid E_1)$ it is clear that

$$P(E_2 \mid E_1) = \frac{P(E_1 \cap E_2)}{P(E_1)}. \tag{1.2}$$

Example 1.4.5. *As an example, we might ask what is the probability we are dealt a pair of aces in the first two cards we receive?*

Solution: Here event E_1 is our first card is an ace. Event E_2 is our second card is an ace. Event E_1 has probability 1/13. However, these two events are dependent. If event E_1 occurs, then it is slightly less likely that E_2 will occur. Here $P(E_2 \mid E_1)$ denotes the probability the second card is an ace, given that the first card is an ace. Thus, as only three aces would remain, we see that $P(E_2 \mid E_1) = 3/51 = 1/17$. Now, we wish both events to happen and so

$$P(E_1 \cap E_2) = P(E_1)P(E_2 \mid E_1) = (1/13)(1/17) = 1/221. \quad \square$$

Another reasonable question is:

Example 1.4.6. *How likely is it that a suited Texas Holdem hand (first two face down cards are of the same suit) will flop (next three face up cards) a flush?*

Solution: Here we have three events in the flop, let E_i be card i is a card of the flush suit, for $i = 1, 2, 3$. But again these are clearly dependent events and so we have

$$P(E_1) = \frac{11}{50}, \qquad P(E_2 \mid E_1) = \frac{10}{49} \text{ and } P(E_3 \mid E_1 \cap E_2) = \frac{9}{48}.$$

Thus, applying Rule 1.4.6 (3) we have

$$P(E_1 \cap E_2) = P(E_1)P(E_2 \mid E_1) = (11/50)(10/49) = 11/245.$$

Now, again applying Rule 1.4.6 (3) we see that

$$P(E_1 \cap E_2 \cap E_3) = P(E_1 \cap E_2)P(E_3 \mid E_1 \cap E_2)$$

$$= (11/245)(9/48)$$

$$= 33/3920$$

$$\approx .0084.$$

Thus, we see that there is a little less than a 1% chance the flop will produce a flush. $\quad \square$

We end this section with another rule that should be obvious.

Rule 1.4.7. Total Probability Formula: *Let A be an event. If the sample space $S = \cup_{i=1}^{n} H_i$ and the H_i are mutually exclusive, then*

$$P(A) = P(A \mid H_1)P(H_1) + P(A \mid H_2)P(H_2) + \ldots + P(A \mid H_n)P(H_n).$$

It is fairly easy to see why this rule holds. The various events (sets) H_i partition the sample space, that is, they have nothing in common (mutually exclusive), but their union is all of the sample space. We can think of $P(A \mid H_i)$ as the proportion of H_i that is a part of A. We obtain this proportion provided we are in H_i already, which happens with probability $P(H_i)$. When we add together all the proportions of A in the various H_i's, we obtain the proportion of the entire sample space that is A, hence, we obtain $P(A)$.

Example 1.4.7. *A baseball team noticed that leadoff hitters (H_1) averaged one strikeout in ten at bats, second place hitters (H_2) averaged one in six at bats, and for the other positions the averages were: H_3 averaged 1/8, H_4 averaged 1/6, H_5 averaged 1/8, H_6 averaged 1/8, H_7 averaged 1/10, H_8 averaged 1/6 and H_9 averaged 1/2. Let K be the event that someone strikes out. What is $P(K)$?*

Solution: Noting that $P(H_i) = 1/9$ for each $i = 1, 2, \ldots, 9$ and using the Total Probability Formula we have that:

$$P(K) = (1/9)\left[1/10 + 1/6 + 1/8 + 1/6 + 1/8 + 1/8 + 1/10 + 1/6 + 1/2\right]$$

$$= (1/9)\left[2/10 + 3/8 + 3/6 + 1/2\right]$$

$$= (1/9)\left[189/120\right]$$

$$= .1750. \quad \square$$

Note that the Total Probability Formula is closely related to Bayes' Formula (see for example [18]).

Rule 1.4.8. Bayes' Formula   *Let A be an event. If the sample space $S = \cup_{i=1}^{n} H_i$ and the H_i are mutually exclusive, then*

$$P(H_i \mid A) = \frac{P(H_i)P(A \mid H_i)}{P(A \mid H_1)P(H_1) + P(A \mid H_2)P(H_2) + \ldots + P(A \mid H_n)P(H_n)},$$

$i = 1, 2, \ldots, n.$

Example 1.4.8. *Suppose that* 10% *of baseball players take steroids. A drug test detects* 90% *of the players taking steroids, but it also falsely identifies* 2% *of the nonusers as taking steroids. What is the probability that a player identified as taking steroids actually is using them?*

Solution: Let S stand for the players taking steroids and let $\overline{S}$ stand for those not taking steroids. We can see that the sample space (the players) are partitioned into two sets by S and $\overline{S}$. Also let U stand for the players identified by the test as using steroids. Then, by Bayes' Formula

$$P(S \mid U) = \frac{P(U \mid S)P(S)}{P(U \mid S)P(S) + P(U \mid \overline{S})P(\overline{S})}$$

$$= \frac{(.9)(.1)}{(.9)(.1) + (.02)(.9)}$$

$$= \frac{.09}{.09 + .018} = .833. \qquad \square$$

Example 1.4.9. *If a football team determined that tight ends* (T) *accounted for* 37% *of the passes caught by receivers, while slot receivers* (S) *accounted for* 42% *and flankers* (F) 21%. *If* 6% *of the passes caught by tight ends are "bombs" (i.e., for* 30 *yards or more) while the corresponding percentage for slot receivers is* 4% *and for flankers is* 12%. *What is the probability that a bomb was caught by a flanker?*

Solution: Let A denote the event that a bomb is caught. Let C_T, C_F and C_S be the events a pass is caught by a tight end, flanker or slot receiver, respectively. Thus, $P(C_T) = .37$, $P(C_S) = .42$ and $P(C_F) = .21$. Further, we see that $P(A \mid C_T) = .06$, $P(A \mid C_S) = .04$ and $P(A \mid C_F) = .12$.

Now using the tree diagram from Figure 1.6 and applying Bayes'

Formula, we have

$$P(C_F \mid A) = \frac{P(C_F)P(A \mid C_F)}{P(C_T)P(A \mid C_T) + P(C_F)P(A \mid C_F) + P(C_S)P(A \mid C_S)}$$

$$= \frac{.0252}{.0222 + .0168 + .0252}$$

$$= \frac{.0252}{.0642}$$

$$= .3925. \qquad \square$$

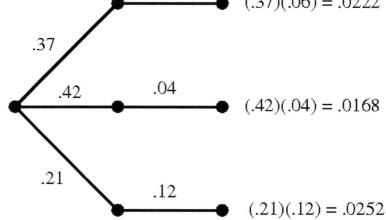

FIGURE 1.6:   The tree diagram for Example 1.4.8.

Example 1.4.10. *A football team noticed that 84% of rushing plays were by halfbacks, 10% by fullbacks, 3% by quarterbacks and 3% by receivers on reverses. They also noticed that for big carries of 10 or more yards, 90% were by halfbacks, 1% by fullbacks, 1% by quarterbacks and 8% by receivers. What is the probability a big carry of 10 or more yards was done by a receiver?*

Solution: Let event A be a rushing play of 10 or more yards. Let R_H, R_F, R_Q and R_R be the events that a rushing play was done by the halfback, fullback, quarterback, or receiver, respectively. We seek the probability $P(R_R|A)$.

Then we know that $P(R_H) = .84$, $P(R_F) = .10$, $P(R_Q) = .03$, and $P(R_R) = .03$. We also know that $P(A|R_H) = .90$, $P(A|R_F) = .01$, $P(A|R_Q) = .01$ and $P(A|R_R) = .08$. Let

$$T = P(R_H)P(A|R_H)+P(R_F)P(A|R_F)+P(R_Q)P(A|R_Q)+P(R_R)P(A|R_R).$$

Now by Bayes' formula we have

$$P(R_R|A) = \frac{P(R_R)P(A|R_R)}{T}$$

$$= \frac{(.03)(.08)}{(.84)(.90) + (.10)(.01) + (.03)(.01) + (.03)(.08)}$$

$$= \frac{.0024}{.756 + .001 + .0003 + .0024}$$

$$= .0032.$$

Exercises:

1.4.1 What is the probability of drawing any 7 or the ace of clubs when drawing one card from a standard deck?

1.4.2 What is the probability of drawing an 8 and a heart in drawing one card from a standard deck?

1.4.3 You flip a coin five times. How large is the sample space of outcomes of the five flips. What is the probability of any one of these equally likely outcomes?

1.4.4 What is the probability of rolling a total of 4 when rolling a pair of dice?

1.4.5 What is the probability of rolling doubles when rolling a pair of dice?

1.4.6 What is the probability of rolling a total of 8 or doubles when rolling a pair of dice?

1.4.7 What is the probability of rolling three 6s when rolling three dice? What about exactly two 6s when rolling three dice? What about exactly one 6 when rolling three dice? What about no 6s when rolling three dice?

1.4.8 A bowl contains 100 coins. One has a head on each side, while the other 99 have a head on one side and a tail on the other side. One coin is picked at random and flipped two times.

1. What is the probability we get two heads?
2. What is the probability we get two tails?

1.4.9 Upon rolling two dice, what is the probability the sum is 8 given that both die show an even number? What is the probability the sum is 12 given both die show an even number? What is the probability the sum is 10 given both die show an odd number?

1.4.10 On a game show you are offered several bowls from which to choose tickets. Bowl I contains three superbowl tickets and seven tickets to the World Series. Bowl II contains six superbowl tickets and four world series tickets. You will select one ticket from one bowl.

1. Compute the probability you selected a superbowl ticket.
2. Given that you selected a superbowl ticket, what is the conditional probability that it was drawn from bowl II?

1.4.11 Police records show that arrests at a certain soccer stadium during a game occur with probability .35. The probability that an arrest and conviction will occur is .14. What is the probability that the person arrested will be convicted?

1.4.12* In the NFL, it is known that the probability is .82 that a first round draft choice who attends all of training camp will have a productive first season and that the corresponding probability for those that do not attend all of camp is .53. If 60% of the first round picks attend all of training camp, what is the probability that a first round pick who had a productive first season will have attended all of camp?

1.4.13* A hockey team noticed that 37% of its goals are scored by centers, 30% by left wings, 20% by right wings and 13% by defensemen. If 20% of the goals scored by centers are power play goals, while 15% scored by wings and 10% scored by defensemen are power play goals:

1. What is the probability that a power play goal was scored by a defenseman?

2. What is the probability it was scored by a left winger?

3. What is the probability it was scored by a right winger?

4. What is the probability it was scored by a center?

1.5 Poker Hands Versus Batting Orders

It should be apparent that to compute probabilities for large events or in large sample spaces, we need to be able to count the number of outcomes that fit our event without having to list them all. This is not always an easy task. In fact, there is an old joke that says, "There are three kinds of mathematicians, those that can count and those that can't."

In order to begin counting in more complicated or larger situations, it will be helpful to recognize the difference between a poker hand and a batting order! In a batting order we have a sequence of items (the hitters) whose position is fixed, that is, the order of their appearance is what matters here. But with a poker hand, it does not matter in what order you receive the cards, all that really matters is the *set of cards* (your hand) you have at the end of dealing. This difference between order mattering and not mattering plays a fundamental role in counting. Let us begin with an example where order matters.

Example 1.5.1. *Suppose your baseball team (of exactly nine players) has a game and you must decide the batting order. How many possible batting orders are there?*

Solution: This is an application of the multiplication principle to the "tasks" of chosing hitters. There are clearly nine choices for the first hitter, then only eight choices for the second hitter, seven choices for the third hitter and so forth. Thus, it should be clear that there are

$$9 \times 8 \times 7 \times \ldots 3 \times 2 \times 1 = 362,880$$

possible batting orders.   □

Each of the different batting orders is called a *permutation*. That is, formally a permutation is an ordered arrangement of the objects from some set. For example, if our set is

$$S = \{1, 2, 3\}$$

then the permutations of the elements of S are:

$$1,2,3 \quad 1,3,2 \quad 2,1,3$$
$$2,3,1 \quad 3,1,2 \quad 3,2,1.$$

We notice that there are six permutations of these three objects. In general, reasoning as we did in the batting order problem, if there are k objects, then the number of permutations on these k objects is:

$$k \times (k-1) \times (k-2) \times \ldots 2 \times 1 = k!$$

where $k!$ is read *k-factorial* . As a matter of convenience we define $0! = 1$.

Next we turn to a problem where order plays no role.

Example 1.5.2. *How many different starting two-card hands are possible in Texas Holdem?*

Solution: There are clearly 52 possible first cards that can be dealt to you and then there are 51 possible second cards. But we do not want to just apply the multiplication principle because that would imply order mattered and for example, count the pair of cards (J ♣, 10♦) and the pair of cards (10♦, J ♣) as two different pairs of cards. But clearly, for the purposes of poker, they are not different pairs because all that matters is the final set of two-cards. Thus, we must adjust for the fact we considered them as ordered!

To do that, first use the multiplication principle and then we divide by 2, to undo the $2! = 2$ possible permutations of the two-cards. Thus, the number of possible two card starting hands in Texas Holdem is

$$\frac{52 \times 51}{2} = 1326. \quad \square$$

In general, this type of counting problem is called a *combination*, that is, a combination is an arrangement of objects in which order does not matter. Poker hands, bridge hands, the toppings on a pizza are all examples of combinations.

In general then, given a set of r objects, we can count the number of combinations of k objects selected from these r objects as follows: use the multiplication principle first and then divide by the number of possible permutations of these k objects:

$$\frac{r \times (r-1) \times \ldots \times (r-k+1)}{k!}.$$

But with a little algebra we can adjust this to be

$$\frac{r \times (r-1) \times \ldots \times (r-k+1) \times (r-k)!}{k! \times (r-k)!} = \frac{r!}{k! \times (r-k)!}.$$

This computation is so common that we denote it as

$$\binom{r}{k} = \frac{r!}{k! \times (r-k)!}. \tag{1.3}$$

The symbol $\binom{r}{k}$ is read as r choose k and is called a *binomial coefficient*. We will see much more about binomial coefficients in Chapter 3.

Now lets consider a few examples of our two kinds of counting.

Example 1.5.3. *How many different five-card poker hands are possible?*

Solution: Because we are dealing with a poker hand, order is not important. What we want to know is how many ways we can choose five cards from a deck of 52 cards. Thus,

$$\binom{52}{5} = \frac{52!}{5!47!} = 2,598,960$$

different five-card poker hands are possible. $\square$

Example 1.5.4. *How many different bridge hands are possible?*

Solution: Again, the final hand is all that matters and the order in which the cards are dealt plays no role. Since a bridge hand consists of 13 cards from a standard deck, our computation is

$$\binom{52}{13} = 635,013,559,600$$

possible bridge hands. (That is surely a lot of possible hands!)   □

Example 1.5.5. *In how many different ways can a bridge player arrange his or her hand?*

Solution: We just saw that a bridge player is dealt 13 cards. We think of this arrangement as an ordering from say, left to right (a permutation). So the number of arrangements of those 13 cards is clearly 13!, that is,

$$13! = 13 \times 12 \times \ldots \times 2 \times 1 = 6,227,020,800. \quad □$$

Example 1.5.6. *How many different foursomes can be formed from among a group of 12 golfers?*

Solution: From the wording of the question, a foursome is a set of four golfers. Thus, this is again a combination problem. That is, the number of foursomes is just

$$\binom{12}{4} = \frac{12!}{4! \times 8!} = \frac{11880}{24} = 495. \quad □$$

Example 1.5.7. *How many possible first, second and third place orders of finish are there in a 10-horse race?*

Solution: Clearly, in this problem, order matters. So we may apply the Multiplication Principle. Thus, there are

$$10 \times 9 \times 8 = 720$$

possible ways for the top three finishers to occur.   □

Permutations and combinations are major counting techniques. We will have occasion to use them many times, especially in the sections on casino games. The reader should become very familiar with each method and comfortable with when to apply them.

Exercises:

1.5.1 List all the permutations of the elements in the set $\{8\spadesuit, 6\clubsuit, A\heartsuit\}$.

1.5.2 How many permutations are possible with a standard deck of 52 cards?

1.5.3 In how many ways can a poker player arrange a five-card hand?

1.5.4 How many three-card poker hands are possible?

1.5.5 In the game of pinochle, each of 4 players are dealt 12 cards from a 48 card deck. How many possible pinochle hands are there?

1.5.6 How many ways are there to arrange a pinochle hand from left to right?

1.5.7 How many ways are there for you and four friends to seat yourselves in five consecutive seats of one row at a basketball game?

1.5.8 Considering the last problem, what if three seats were in one row and two seats in the next row?

1.5.9 If three couples go to the Superbowl and have six consecutive seats in one row, how many ways are there to arrange the seating if:

1. Couples must be seated together?
2. The men must be seated together?
3. Couples must not be seated together?

1.5.10 Your basketball team has 12 players: 5 guards, 5 forwards and 2 centers. If you are going to start 2 guards, 2 forwards and a center, how many starting lineups are possible?

1.5.11 As in the previous problem, suppose you are going to play a "small" lineup of 3 guards, 1 forward and 1 center. Now how many starting lineups are possible? What if you start 3 guards and 2 forwards?

1.5.12 If you hold 3 clubs, 4 spades, 2 hearts and 4 diamonds and you keep cards of the same suit together, how many ways are there to arrange your hand so that

1. The suits from left to right are clubs, diamonds, hearts and spades?

2. If the order of the suits does not matter?

3. If the suits are red, black, red, black (left to right)?

1.6   Let's Play for Money!

Those words should always cause you to be suspicious. Anyone who wants to play a game for money also wants to make a profit at the game. In order not to be at a (potentially serious) disadvantage, we need to make some decisions as to how "fair" it is to play the game. Remember the warning from Cardano! In this section we develop one fundamental way of deciding this question.

Let us consider a simple example. How much would you be willing to pay to play the following game?

A single coin is tossed and we count how many tosses it takes before the first head appears. This might take one try and it might take many tries. Here we limit you to three tries. Based on the rules, the sample space for this game is

$$S = \{H, TH, TTH, TTT\}.$$

We use the payments shown in Table 1.3.

To answer our question of how much you might be willing to pay in order to play this game, we want to consider the average amount we would win. If we know the average amount we will win upon playing this game, we should be willing to spend up to that amount to play.

To determine the average amount we win at this game we need to determine the probabilities of the various outcomes. These are given in Table 1.4.

TABLE 1.3: Payment Table for the Coin Flip Game

H	TH	TTH	TTT
$1	$2	$4	$0

TABLE 1.4: Probabilities of the Outcomes in S

H	$\frac{1}{2}$
TH	$\frac{1}{2} \times \frac{1}{2} = \frac{1}{4}$
TTH	$\frac{1}{2} \times \frac{1}{2} \times \frac{1}{2} = \frac{1}{8}$
TTT	$1 - (1/2 + 1/4 + 1/8) = 1/8$

We can see that based on these probabilities and the corresponding payments, our expected average winning would be:

$$(1/2) \times \$1 + (1/4) \times \$2 + (1/8) \times \$4 + (1/8) \times \$0 = \$1.50.$$

Thus, it appears we should be willing to pay up to $1.50 to play this game. If we pay more, we would expect to lose money in the long run (by repeatedly playing the game) as our average rate of return is below the cost of playing. While if we pay less we would expect to win money in the long run as our average rate of return would exceed the cost of playing.

This is a typical question and this probability weighted "average" is called the *expected value*. Expected values can be computed for many different things, not just for payments. To formalize this idea we need another definition.

Given a random experiment with a sample space S, a function X which assigns to each element $s \in S$ one and only one real number $X(s) = x$ is called a *random variable*. Thus, for example, in a dice game we may have one random variable X representing the payouts for the possible game outcomes and another random variable Y representing the possible rolls of the die. We might even have one random variable that models the roll of one die and another random variable for the other die.

With the idea of random variables in mind we can now define expected value formally. Let X be a random variable and let $e_1, e_2, \ldots, e_k$ be the possible outcomes of an experiment with corresponding probabilities $p_1, p_2, \ldots, p_k$. Then the expected value (EV) of this random variable is:

$$EV(X) = p_1 \times X(e_1) + p_2 \times X(e_2) + \ldots + p_k \times X(e_k).$$

That is, an expected value is a probability weighted average of the values taken by the random variable. If all $p_i = 1/k$, then we really have an average of the values of the random variable.

Thus, from our example above, the expected value of the coin flip game is \$1.50. If we pay \$1.50 to play, our expected value computation for playing the game then becomes:

$$(1/2) \times \$1 + (1/4) \times \$2 + (1/8) \times \$4 + (1/8) \times \$0 - \$1.50 = \$0.00.$$

We use an expected value of zero as an indicator of a *fair game*. There is no clear advantage to either side.

Expected values can be computed for many different experiments and do not necessarily involve money.

Example 1.6.1. *What is the expected value of the roll of a single die?*

Solution: Here the expected value computation becomes straightforward since the (uniform) probability of each outcome is $1/6$, our random variable X is the value rolled, we have:

$$EV(X) = (1/6)[1 + 2 + 3 + 4 + 5 + 6]$$

$$= (1/6)(21) = 3.5. \quad \square$$

Thus, the expected roll of a single die is 3.5. But we cannot actually roll 3.5. This example is important as it shows that the expected value is really an average and not necessarily even a possible outcome.

Exercises:

1.6.1 Find the expected value for the game of Exercise 1.3.7.

1.6.2 Find the expected value for the sum when we roll a pair of dice.

1.6.3 Below are the probabilities for the USC football team winning n games in the first half season. What is the expected number of games USC will win in the first half season?

n	0	1	2	3	4	5	6
P(n)	.001	.010	.060	.185	.304	.332	.108

1.6.4 You play a game where the probability of winning is .45. You win $3 if you win the game and lose $2 otherwise. What is the expected value of playing this game?

1.6.5 In the game from the previous problem, suppose instead that you would win or lose $1. Now what is the expected value of playing this game?

1.6.6 Odds makers try to predict which football team will win and by how much (the *spread*). If they are correct, adding the spread to the loser's score would produce a tie. Suppose you can win $6 for every dollar you bet if you can predict the winner of three consecutive games. What is your expected value for this bet?

1.6.7 If the random variable X assigns to each card of a standard deck the face value of that card, except $X(ace) = 1$ and $X(jack) = X(queen) = X(king) = 10$. What is the expected value of X?

1.6.8* You play a game with probability p that you will win. You will play until you either win, or have lost three consecutive times. What is the expected number of times you will play this game? (Hint: Think of the sample space and the associated probabilities.)

1.6.9* There are five marbles in a jar. Three of the marbles are red and two are blue. What is the expected number of times you must randomly select a marble in order to select a blue marble, assuming that you do not replace selected marbles? What if you do replace the selected marbles?

1.7   Is That Fair?

In this section we wish to consider some other examples of the use of probabilities, expected values and in general, ideas about fairness in games. Since we have not given fairness a mathematical definition, we have many possible ways to interpret what we might mean. Our examples are intended to show there are many ways to determine fairness and many subtle ways to tip the scales in your own favor.

Our first application is in the idea of fair division of a prize. We consider the problem mentioned earlier that intrigued Pascal and Fermat, called **The Problem of Points.** We demonstrate the idea behind this problem with an example.

Example 1.7.1. *Suppose Jane and Tom are playing a game that requires the winner to reach a total of 5 points. Jane is leading 4 points to 2 when the game is forced to halt. How shall we fairly divide the prize money for this unfinished game?*

Solution: Our proposed solution is to divide the prize money based on the probability of winning for each player. Since the players need to reach 5 points to win, a total of 9 points are possible (5 for one and 4 for the other). We shall project the possible play forward to 9 points, assuming each outcome is equally likely, to see the chances for each player. Can you determine how many possible sequences we must consider?

TABLE 1.5:   The Possible Outcomes with Further Play

Point 7	Point 8	Point 9	Winner
J	J	J	J
J	J	T	J
J	T	J	J
T	J	J	J
T	T	J	J
T	J	T	J
J	T	T	J
T	T	T	T

Now, of the eight possible ending scenarios to the game (see Table 1.5), Jane wins in seven of them. It would seem fair to give Jane 7/8 of the prize money and Tom just 1/8. □

The general Problem of Points asked for a solution to this type of problem, no matter what the partial score or the final winning score. The process above lends itself to a solution of this general problem. However, we shall not go through the details for the solution to the general problem.

Next let's take a look at another expected value problem often called the *St. Petersburg Paradox*. As you might expect from the name, this question raises a big issue.

Example 1.7.2. The St. Petersburg Paradox. *A fair coin is tossed repeatedly until a head comes up. If it comes up heads on toss one, you are paid $2. If it first comes up heads on toss two, you are paid $4. This continues in that if the first head appears on the i-th toss, then you are paid $X(i) = \$2^i$. What is the expected value of this game?*

Solution: Note that this is an unlimited version of the three-toss game we considered in the previous section. The probability of a head on toss one is $\frac{1}{2}$. In general, the probability of the first head on the i-th toss is $\frac{1}{2^i}$. Thus, the expected value of this game is:

$$EV(X) = \sum_{n=1}^{\infty} \frac{1}{2^n} 2^n = \sum_{n=1}^{\infty} 1 = \infty.$$

Therefore, arguing as we did before, the fair price to pay to play this game is an INFINITE number of dollars! But of course, no one can or would pay that amount. To recover your investment would mean winning when the first head did not appear until 2^n was at least the amount invested, so you would need the head not to appear for an infinite number of tosses. But this would happen only with a very tiny (infinitesimal) probability.

Thus, the moral of the story here is that expected value can be used as a fairness indicator, but only when the game in question can actually be played often enough to allow the player a chance to win back the price! In a game that is only going to be played once (or very few times), it is not necessarily a good guide. A game requiring an infinite number of plays of course makes things impossible. The point we need to take away from all this is that expectation (and any probability argument in general) is meaningful only for long-term play where there is a real possibility for the long-term play. The experiment must be repeated over and over again for the expected value to really have significance, and the number of repetitions must be possible.   □

Now let's take a different look at "fairness."

Example 1.7.3. The game of Penny-Ante. *A fair coin is to be tossed repeatedly. The outcomes are recorded as H for a head and T for a tail. You allow Jane to select any pattern of three outcomes she wishes and from the remaining patterns of three outcomes you select one of your own. The game is now to flip a coin until one of the selected patterns occurs. The first player to have his pattern occur is the winner.*

Solution: This game was introduced by Walter Penny in 1969 (see [16]). Below is the sequence of outcomes when I actually tossed a coin 20 times. I have grouped them in blocks of 5 for convenience.

$$HTHHT \quad HHHHT \quad HTTHH \quad THHTT$$

Penny-Ante is based on patterns of length three. The first such pattern in the sequence I created is HTH, using positions $1, 2, 3$, and the second pattern uses positions $2, 3, 4$ and is THH and so forth. Clearly there are $2^3 = 8$ possible patterns as any of the three positions can be either of the two possible outcomes.

  The game may appear slightly unfair. It might seem that Jane has a slight advantage in picking first as she may select any pattern and we do not have that option. In reality, the game is unfair, but not for this reason and not in Jane's favor! The fact is that picking second allows you to select a pattern that will always have at least a $2/3$ chance of winning!

  One way to see the "scam" here is to suppose that Jane chooses HHH as her pattern. If HHH comes up on the first three tosses then Jane wins. This should happen approximately $1/8$ of the time. For the remaining $7/8$ of the time, this fails to happen. Thus, a T must occur within the first three tosses. Perhaps now you can see the advantage of selecting the pattern THH. Before the pattern HHH can first occur somewhere other than in the first three tosses, the pattern THH would already have to occur! Thus $7/8$ of the time your pattern

TABLE 1.6: The Patterns, Some Responses
and Probabilities

Jane's Choice	Your Choice	Prob. You Win
HHH	THH	7/8
HHT	THH	3/4
HTH	HHT	2/3
THH	TTH	2/3
HTT	HHT	2/3
THT	TTH	2/3
TTH	HTT	3/4
TTT	HTT	7/8

will win. A table of the patterns, what you should select in response
to your opponent's choice of pattern and the probability you will win,
are listed in Table 1.6. Note the symmetry of probabilities from top to
bottom of the table.

Figure 1.7 shows the relationships between the choices, with the arrows pointing at the superior choice. It is interesting to note the inner
cycle, where one choice dominates another, which in turn is dominated
by another, which in turn is dominated by the fourth choice, which
again dominates the first. An easy way to remember how to choose
your sequence is to move the first two choices of your opponent to your
second and third positions and never pick a palindrome.

What about the other cases. Here we will show how to attack the
more complicated cases. Suppose Jane has selected the pattern *HTT*
and I have selected the pattern *HHT*. There are many ways for this
to occur and we shall use a *tree diagram* as a means of tracking all
the cases. In the tree diagram, we take a right branch whenever a
heads (H) occurs, and a left branch whenever a tails (T) occurs. The
beginning point of the tree is called its root and is labeled R. Consider
the overview in Figure 1.8. This gives an overview of the key parts of
the tree diagram. The triangles T_1, T_2, etc. represent subtrees of the
diagram that we shall consider in more detail. The subtree T_1 is shown

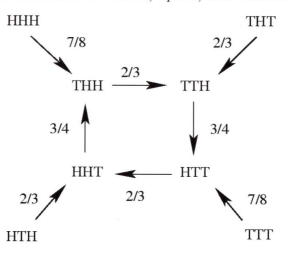

FIGURE 1.7:   The dominance relations for Penny Ante.

in Figure 1.9.

What we see in subtree T_1 is that there is an infinite sequence of possible winning moves for me. We can determine the probabilities for reaching each of these positions as being just $(\frac{1}{2})^i$, where there are i edges on the path from the root of the tree to the winning position. Thus, within the tree T_1 there is a probability of

$$\frac{1}{8} + \frac{1}{16} + \frac{1}{32} + \dots$$

for my winning the game.

What we notice here is that each term of this sum is one-half the previous term. This is an example of a *geometric series* and there is a fixed formula for computing the sum of such a series.

Theorem 1.7.1. *Given a sum*

$$a + ar + ar^2 + ar^3 + \dots + ar^{n-1} + \dots$$

where r is a fixed number less than 1, then the sum of these values is given by

$$\frac{a}{1-r}.$$

Note, the proof of this result is beyond the scope of this book, hence we shall just accept it.

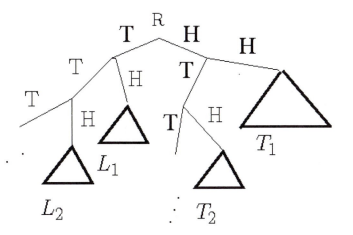

FIGURE 1.8:   The general tree diagram.

Now, considering our sum again and applying the theorem we see that

$$\frac{1}{8} + \frac{1}{16} + \frac{1}{32} + \ldots = \frac{1/8}{1 - 1/2} = \frac{1}{4}.$$

Next we note that the subtree T_2 is similar to T_1, except each value is multiplied by an additional $1/4$. Hence T_2 will contribute $1/16$ to the probability we win, then T_3 will contribute $1/64$, etc.

Hence, all the subtrees from the right-hand side of the tree contribute

$$\frac{1}{4} + \frac{1}{16} + \frac{1}{64} + \ldots = \frac{1}{4} \times \frac{1}{1 - 1/4} = \frac{1}{3}.$$

to the probability we win.

Now each of the subtrees on the left-hand side is similar to the entire right-hand side, except each has been multiplied by an extra $1/2$, for each extra edge of the tree we must traverse in order to reach the subtree. Thus, L_1 contributes $1/6$ to the probability we win, while L_2 contributes $1/12$, etc. Combining all these subtrees we have that the probability I win is

$$\frac{1}{3} + \frac{1}{6} + \frac{1}{12} + \ldots = \frac{1}{3} \times \frac{1}{1 - 1/2} = \frac{2}{3}.$$

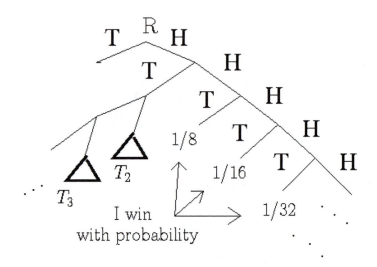

FIGURE 1.9:   The right subtree in more detail.

Thus, as long as I pick second, a similar argument will show that I can always have a probability of at least 2/3 of winning the game. □

Exercises:

1.7.1 Jane will repeatedly toss a coin until it comes up heads. If the coin comes up heads on the j-th toss where $j \leq 5$, Tom will pay Jane 2^j dollars. But if the coin never comes up heads in five tosses, Tom pays Jane nothing. What is the expected amount Jane will receive from Tom?

1.7.2 In the game above, what is the probability the game will last longer than three tosses? Longer than four tosses?

1.7.3 In the game from Exercise 1.7.1, what is the probability that a head will not come up at all?

1.7.4 Jane and Tom are playing a game to 7 points and presently Tom has 6 points and Jane has 5 points. The game must be halted. How should Tom and Jane divide the prize money?

1.7.5 In the previous problem, determine the number of sequences needed to determine how to divide the prize money without listing them.

1.7.6 Suppose that in Problem 1.7.4 Tom has 5 points and Jane has 4 points. Now how should the money be divided?

1.7.7 Next suppose that Tom has 8 points and Jane has 7 points in a game to 11 points. Now how should the money be divided?

1.7.8 This time Jane has 9 points and Tom has 8 points in a game to 11 points. Now how should the money be divided?

1.7.9* Another dice game (see [16]) has four dice numbered as shown below.

$$\text{Red Die: } 4,4,4,4,4,4 \qquad \text{Blue Die: } 8,8,2,2,2,2$$
$$\text{Green Die: } 7,7,7,1,1,1 \qquad \text{Yellow Die: } 6,6,6,6,0,0$$

Tom and Jane each select one die. The game is simple, they each roll their die once and the winner is the one with the highest value rolled. Show that whoever picks second in this game has a probability of at least 2/3 of winning.

1.7.10* A man has three cards, one red on both sides, one black on both sides, and one red on one side and black on the other. He places the three cards into a hat and asks you to blindly select one card and only look at one side of the card. If that side is red (black), he offers to bet you even money the card is the red-red card (black-black card). Is this a good bet?

1.7.11 Show that if Penny Ante is played with patterns of length two there are choices where the game is fair.

1.8   The Odds Are against Us

We next explore the mathematical meaning of odds. We hear all the time that the odds are against us, or that this horse is the odds on favorite, or that the odds against the Red Sox winning the World

Series this year are presently listed as 8 to 1. We wish to clarify the exact meaning here.

Given an event E (like the Red Sox winning the World Series), we may ask what are the odds against event E happening? We state this as *odds against the event*, as that is the most likely way you will actually see odds stated. The *odds* against event E are defined to be:

$$\text{odds against E} = \frac{P(\overline{E})}{P(E)} = \frac{1 - P(E)}{P(E)}.$$

That is, the odds against event E happening are simply the ratio of the probability E fails to the probability E occurs. For example, if the probability you will win a game of chess against a certain grand master you are about to play is $2/15$, then the odds against you are:

$$\frac{13/15}{2/15} = \frac{13}{2}.$$

We ordinarily try to write the odds in the form of $m{:}n$ where m and n are integers, usually in lowest terms, and we read this as the odds are m to n. Thus, in our example above, the odds are 13:2.

Example 1.8.1. *What are the odds against rolling a total of 7 or 11 with one roll of a pair of dice?*

Solution: Using Table 1.1 we see that there are six ways to roll a 7 and there are two ways to roll an 11. Thus our event E, rolling a 7 or 11 has 8 possible positive outcomes and 28 negative outcomes. Thus, $P(E) = 8/36$ and $P(\overline{E}) = 28/36$. Hence,

$$\text{odds against } E = \frac{28/36}{8/36} = 28/8 = 7/2.$$

Hence, the odds against rolling a 7 or 11 are 7:2.   □

Example 1.8.2. *What are the odds against drawing an Ace from a standard deck?*

Solution: There are four aces in a standard deck and so $P(\text{ace}) = 4/52$ and then $P(\text{no ace}) = 1 - P(\text{ace}) = 48/52$. Hence,

$$\frac{48/52}{4/52} = \frac{12}{1}$$

and so the odds against drawing an ace are 12:1.   □

Exercises:

1.8.1 What are the odds against selecting a spade when you draw one card from a standard deck of 52 cards?

1.8.2 How would you define *the odds for* an event E?

1.8.3 What are the odds against rolling 12 when you roll a pair of dice?

1.8.4 What are the odds against rolling three 6s when you roll three dice?

1.8.5 What are the odds against being dealt two aces when you are dealt two cards from a standard deck?

1.8.6 What are the odds against drawing a face card (i.e., J, Q, K) when you draw one card from a standard deck?

1.8.7 What are the odds against rolling an even number when you roll one die? What about two dice?

1.8.8 What are the odds against rolling doubles when you roll two dice?

1.8.9 Using Example 1.4.5, what are the odds against flopping a flush when you hold two cards of the same suit in Texas Holdem?

1.8.10 Using Example 1.4.4 as a base, what are the odds against being dealt a pair on the first two cards in Texas Holdem?

1.9   Things Vary

We continue this chapter with a look at some of the most common measures in statistics. These measures are used to describe distributions, both of probabilities and of other data sets. They will help us answer natural questions about these distributions.

Given a *discrete* (that is, has only finitely many values) random variable X, the *mean* of X, denoted μ is just the expected value of X. The physical analogy here is to the center of gravity. The mean is the "center" of the distribution in the sense of probability weighting rather than mass of objects. Since all our examples will be discrete, this idea carries nothing really new for us other than the common terminology.

Now that we have a center of our distribution, we can measure the degree of dispersion of the distribution around the center; that is, we want to measure how scattered the points are with respect to this center. This measure is called the *variance*. Formally, the variance of a discrete random variable X, denoted by $var(X)$, is defined to be the expected value of $[X - EV(X)]^2$; that is,

$$var(X) = EV([X - EV(X)]^2). \tag{1.4}$$

Roughly, if the dispersion of X about its mean is small, that is many values of X are close to μ, then $|X - \mu|$ (that is, $|X - EV(X)|$) tends to be small, and so $var(X) = EV(|X - \mu|^2)$ is also small. On the other hand, if the dispersion of X about its mean is considerable, then $|X - \mu|$ tends to be large, giving us that $var(X)$ is also large.

There are different ways in which one might compute the variance. We let $Im(X)$ denote the *image* of the random variable X; that is, the values taken on by X. For example, if X is the random variable representing the roll of a single die, then

$$Im(X) = \{1, 2, 3, 4, 5, 6\}.$$

Now, for a discrete random variable X,

$$var(X) = EV([X - EV(X)]^2) = \sum_{x \in Im(X)} (x - \mu)^2 P(X = x) \tag{1.5}$$

where

$$\mu = EV(X) = \sum_{x \in Im(X)} x P(X = x).$$

Equation 1.5 is not always the easiest way to compute the variance of a discrete random variable. We may expand the term $(x - \mu)^2$ to obtain

$$var(X) = \sum_{x \in Im(X)} (x^2 - 2x\mu + \mu^2)P(X = x)$$

$$= \sum_{x \in Im(X)} x^2 P(x) - 2\mu \sum_{x \in Im(X)} xP(x)$$

$$+\mu^2 \sum_{x} P(x)$$

$$= EV(X^2) - 2\mu^2 + \mu^2$$

$$= EV(X^2) - \mu^2$$

where again $\mu = EV(X)$ as before. Hence we obtain the following useful formula:

$$var(X) = EV(X^2) - EV(X)^2. \tag{1.6}$$

As a measure of dispersion, variance has one undesirable property, it is not a linear function. That is, the variance of aX is $a^2 var(X)$, for any real number a. For this reason, statisticians often prefer to work with the *standard deviation* of X, denoted $\sigma(X)$, defined to be

$$\sigma(X) = \sqrt{var(X)}.$$

The reader should note that $var(X)$ is often denoted $\sigma^2(X)$ because of this relationship. We shall recognize both notations.

Example 1.9.1. *Suppose the random variable X takes on the value 1 when a head is flipped and -1 when a tail is flipped. What is the mean, variance and standard deviation of X?*

Solution: For the random variable X, the image $Im(X) = \{1, -1\}$ and so the mean is clearly

$$\mu = 1(1/2) + (-1)(1/2) = 0.$$

Thus, computing the variance we have

$$var(X) = P(1)(1 - \mu)^2 + P(-1)(-1 - \mu)^2$$
$$= 1/2(1 - 0)^2 + 1/2(-1 - 0)^2$$
$$= 1.$$

Then clearly, the standard deviation is also 1. $\qquad\qquad\square$

Example 1.9.2. *What is the mean, variance and standard deviation for rolling one die?*

Solution: Let X be the random variable for the outcomes of rolling one die. Earlier, in Example 1.6.1 we determined that the expected value of rolling one die is $\mu(X) = 3.5$. Thus, the variance of X is computed as

$$var(X) = \sum_{x \in Im(X)} P(X = x)(x - \mu)^2$$

$$= 1/6[(1 - 3.5)^2 + (2 - 3.5)^2 + (3 - 3.5)^2$$

$$+ (4 - 3.5)^2 + (5 - 3.5)^2 + (6 - 3.5)^2]$$

$$= 1/6(6.25 + 2.25 + .25 + .25 + 2.25 + 6.25)$$

$$= 1/6(17.5)$$

$$= 2.91.$$

Now, the standard deviation is $\sigma(X) = \sqrt{2.91} = 1.71.$ □

Exercises:

1.9.1 Suppose you are playing the following dice game. You roll one die and win $5 if a 6 is rolled, and lose $1 otherwise. What is the expected value (mean), variance and standard deviation for this game?

1.9.2 Suppose we have the following random variable X: The value of X is 1 for (H, H), 2 for (H, T), 3 for (T, H) and 4 for (T, T). What is the mean, variance and standard deviation for X?

1.9.3 What is the mean, variance and standard deviation for the random variable representing the sum of two dice?

1.9.4* Suppose the random variable X has the value of the card selected from a standard deck, where an ace counts 1, a jack as 11, queen as 12 and king as 13, with all other cards counting as their face value. What is the mean and variance for X?

1.9.5 Find the variance and standard deviation for the probability distribution of Exercise 1.6.3.

1.9.6 Find the variance and standard deviation for the distribution of Exercise 1.6.4.

1.9.7* Find the variance and standard deviation for the distribution of Exercise 1.7.1.

1.10 Conditional Expectation

In this section we look at expected values when we are dealing with conditional probabilities. It should be reassuring that things work very much the same way.

Let X be a discrete random variable and let E be an event with $P(E) > 0$. Then the *conditional expectation* of X given E, denoted $EV(X|E)$, is defined to be

$$EV(X \mid E) = \sum_{x \in ImX} xP(X = x \mid E).$$

The next result is a natural one in view of the definitions.

Theorem 1.10.1. *If X is a discrete random variable and $\{E_1, \ldots, E_n\}$ is a partition of the sample space such that $P(E_i) > 0$ for each i, then*

$$EV(X) = \sum_i EV(X \mid E_i)P(E_i).$$

Proof. From our definition of conditional expectation we see that

$$\sum_i \sum_x xP(\{X = x\} \cap E_i) = \sum_x xP(\{X = x\} \cap (\cup_i E_i))$$

$$= \sum_x xP(X = x). \quad \square$$

Example 1.10.1. *A coin is tossed repeatedly and heads appears with probability p where $0 < p = 1 - q < 1$. Find the expected length of the initial run (this is a run of heads if the first toss is a head, and a run of tails otherwise).*

Solution: Let E be the event that the first toss is a heads and let $\overline{E}$ be the event the first toss is a tails. The pair E, $\overline{E}$ forms a partition of the sample space. Let N be the length of the initial run. Then we see that

$$P(N = k \mid E) = p^{k-1}q \text{ for } k = 1, 2, \ldots$$

since if E occurs, then N equals k if, and only if, the first toss is followed by exactly $k-1$ more heads and then a tail. A similar argument shows

$$P(N = k \mid \overline{E}) = q^{k-1}p \text{ for } k = 1, 2, \ldots.$$

Thus, by the definition of $EV(X \mid E)$ we see that

$$\begin{aligned}
EV(N \mid E) &= \sum_{k=1}^{\infty} kp^{k-1}q \\
&= \sum_{k=1}^{\infty} kp^{k-1}(1-p) \\
&= \sum_{k=1}^{\infty} kp^{k-1} - \sum_{k=1}^{\infty} kp^k \\
&= S_1 - S_2.
\end{aligned}$$

Now consider the two sums

$$\begin{aligned}
S_1 &= 1 + 2p + 3p^2 + 4p^3 + \ldots \\
S_2 &= -p - 2p^2 - 3P^3 - \ldots
\end{aligned}$$

so that,

$$S_1 - S_2 = 1 + p + p^2 + p^3 + \ldots$$

and since $p < 1$, we have a geometric series. Applying Theorem 1.7.1 we see that

$$\sum_{k=1}^{\infty} kp^{k-1}q = \frac{1}{1-p} = \frac{1}{q}.$$

and similarly,

$$EV(N \mid \overline{E}) = \frac{1}{p}.$$

Now by Theorem 1.10.1 we see that

$$EV(N) = EV(N \mid E)P(E) + EV(N|\overline{E})P(\overline{E})$$
$$= \frac{1}{q}(p) + \frac{1}{p}(q)$$
$$= \frac{p^2 + 2pq + q^2}{pq} - \frac{2pq}{pq}$$
$$= \frac{(p+q)^2}{pq} - 2$$
$$= \frac{1}{pq} - 2. \quad \square$$

We note that if $p = q = \frac{1}{2}$, the $EV(N) = 2$.

Exercises:

1.10.1 Let N be the number of tosses of a fair coin up to and including the appearance of the first head. By conditioning on the first toss, show that $EV(N) = 2$.

1.10.2* The probability of obtaining a head when a certain coin is tossed is p. The coin is tossed repeatedly until exactly n heads in a row appear. Let N be the total number of tosses required for this to happen. Find the expected value of N.

Chapter 2

The Game's Afoot

2.1 Applications to Games

In the Chapter we wish to apply some of the things we have already learned. We shall use a variety of games as models. This will allow us to hone our skills as well as see how useful these mathematical ideas can be in evaluating these games. We will stress counting techniques, basic probability, conditional probability and expected values. The games considered will include roulette, craps, poker, backgammon and an old television show called *Let's Make A Deal*. Hopefully there will be something of interest in this list for the reader. We begin with poker.

2.2 Counting and Probability in Poker Hands

In this section we wish to apply some of the counting techniques we learned in the last chapter. In particular, we wish to begin by counting the number of ways a five-card poker hand can occur. The reader hoping for a deeper exploration of the mathematics of poker should see [13] and [11]. Let's begin at the start.

Poker is played with a standard 52-card deck. Each card has two attributes, a *rank* and a *suit*. The rank of a card can be any of thirteen possibilities:

$$\{2, 3, 4, 5, 6, 7, 8, 9, 10, J, Q, K, A\},$$

while the suit can be any of four possibilities:

$$\{\clubsuit, \diamondsuit, \heartsuit, \spadesuit\}.$$

We exclude jokers and wild cards.

Example 2.2.1. *The number of different five-card poker hands was an earlier example.*

Solution: This is a standard combinations question and the solution was $\binom{52}{5} = 2,598,960$. □

Example 2.2.2. Counting straight flushes:

Solution: A *straight flush* is a hand with five consecutive ranks, all of which are of the same suit. Straights may begin with an ace, two, ... , or 10. Thus there are $\binom{10}{1}$ ways to begin the straight with the remaining cards all being fixed. There are also $\binom{4}{1}$ ways to select the suit for the straight flush. Thus, by the Multiplication Principle there are a total of $\binom{10}{1}\binom{4}{1} = 40$ possible straight flushes. Note that 4 of these are the so-called *royal flushes* $(10, J, Q, K, A)$. □

Example 2.2.3. Counting four of a kind:

Solution: This is four cards of the same rank. The fifth card is necessarily of a different rank and can be any such card. Our computation is then: $\binom{13}{1}$ ways to select the rank, and $\binom{4}{4}$ ways to select the four cards of that rank. Now the last card can be selected in $\binom{48}{1}$ ways, as any card of another rank will do. Thus, by the multiplication principle the number of four of a kind hands is:

$$\binom{13}{1}\binom{4}{4}\binom{48}{1} = 624. \quad \square$$

Example 2.2.4. Counting full houses:

Solution: A *full house* consists of three cards of one rank and two cards of another rank. There are $\binom{13}{1}$ ways to select the first rank and $\binom{4}{3}$ ways to select the cards within that rank. Then there are $\binom{12}{1}$ ways to select the second rank and $\binom{4}{2}$ ways to select the two cards of that rank. Thus, the total number of such hands is:

$$\binom{13}{1}\binom{4}{3}\binom{12}{1}\binom{4}{2} = 3744. \quad \square$$

TABLE 2.1: Counting Five-Card Poker Hands

HAND	HOW TO COUNT	NUMBER
Straight Flush	$\binom{10}{1}\binom{4}{1}$	40
Four of a kind	$\binom{13}{1}\binom{48}{1}$	624
Full house	$\binom{13}{1}\binom{4}{3}\binom{12}{1}\binom{4}{2}$	3744
Flush		
Straight		
Three of a kind		
Two Pair		
One pair		
High card		

Problem 2.2.1. *Complete Table 2.1.*

Example 2.2.5. *Suppose you hold the 7 of spades and the 7 of hearts in a Texas Holdem game. The flop is king of clubs, 8 of diamonds, 3 of hearts. What is the probability you will make three 7s on the next card?*

Solution: We shall ignore the burn cards (cards removed before the flop, turn and river, etc.) and as we cannot determine what is in our opponent's hands, we also ignore these cards in answering this question. There are only two 7s remaining in the deck. Thus, there are five cards we know and 47 we do not know. Now we conclude that the probability the next card is a 7 is $2/47 = .0426$. That is, we have a 4.27% chance of getting the desired card.   □

Note that we assumed all of the cards that could help you were still in the deck. This might not be true, so our number is really an upper bound on your chances of making the desired hand.

Example 2.2.6. *Suppose you hold two aces, two jacks and a 7 in a five-card poker game with five other players. Upon replacing one card, what is the probability that you will make a full house?*

Solution: Again we see only five cards. We consider all others in play. Thus, 47 cards remain, and only four of those can possibly produce a full house. Thus, the probability you make a full house is $4/47 = .0851$.   □

Note to players: A simple way to estimate the probability of hitting the card you want on the next card in a Texas Holdem game is to count the number of cards that will help you (as we did above with the two 7s), then double that number and add one. So if we had done that in Example 2.2.5 above, we would have estimated we had a 5% chance while our computation said 4.27%. Thus, it was a very good estimate. The idea is there are approximately 50 cards left (always less) so doubling the count takes it approximately 100 cards, so this total tells us the percentage. But then we add one to this total to help offset the overestimate of the cards left in the deck.

One final note is that the ranks of the hands in poker correspond exactly to the frequency with which the hand can appear. The fewer ways the hand can appear, the higher the rank of the hand. Thus, the ranking of poker hands makes perfect mathematical sense!

Exercises:

2.2.1 You are dealt a $4, 7, 8, 9, 10$ in five-card draw poker. What is the probability of drawing one card and making a straight?

2.2.2 What is the probability of getting a flush in the first five cards of a poker hand?

2.2.3 What is the probability of being dealt a straight in a five-card poker hand?

2.2.4 How many three-card poker hands are possible?

2.2.5 What is the probability of a flush in three-card poker?

2.2.6 You are dealt four diamonds and the king of spades. What is the probability of drawing one card and making your flush? (Here ignore the cards of other players as you do not know what they are.)

2.2.7 Complete the following table similar to Table 2.1, but this time for three-card poker hands.

TABLE 2.2:   Counting Three-Card Poker Hands

HAND	HOW TO COUNT HANDS	NUMBER
Straight flush		
Three of a kind		
Straight		720
Flush		
One pair		3744

2.2.8 You hold a $3, 4, 6$ and 7, as well as a 10. What is the probability of drawing one card and making a straight?

2.2.9 Estimate the probability of making a flush on the river (the fifth face up card) in Texas Holdem, provided you have four spades already (two in your hand and two face up on the board)?

2.3   Roulette

History tells us that *roulette* began in France in the 18th century. The game has been played essentially in its current form since around 1796. In 1842 Louis Blanc added the "0" to the roulette wheel in order to gain a house advantage (this is called *European Roulette* as it is still played with one "0" in Europe). In the early 1800s the game migrated to the United States where a "00" was added to further increase the house advantage (see [8]). We call this version *American Roulette*. The game remains more popular in Europe than in the United States, although it is played in every casino in the country.

The game of roulette is simple to play. There is a wheel with the numbers 1 to 36 as well as 0 and 00 each represented by a slot. See Figure 2.1 for a typical American roulette wheel. Half the numbers from 1 to 36 are red, the others black (see Figure 2.2 where black numbers are more darkly shaded than the red or green numbers). Both 0 and 00 are green. A ball is started around the outside of the wheel while the slots spin in the opposite direction. Gradually the ball falls into one of the slots, determining the number and deciding which of the possible bets wins.

Bets may be placed on a variety of possibilities such as the number itself, whether the number was red or black, odd or even and many more possibilities shown in Table 2.3 and Figure 2.2 (which shows a typical roulette table).

The place where the House (casino) makes money is based on the difference between the true odds for the winning event and the *House odds*, that is, the odds actually paid by the casino. We shall now consider several examples to see the effects of these differences.

Example 2.3.1. *What is the expected value of a $1 bet on the number 27 in American Roulette?*

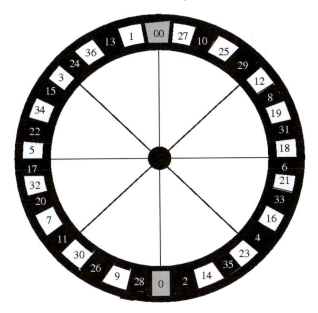

FIGURE 2.1: American Roulette wheel.

Solution: The computation is straightforward. The probability that 27 will be the number is clearly 1/38. Let X denote the random variable for the amount won or lost. The House odds on a single number bet are 35:1 and thus $X(27) = 35$ and $X = -1$ otherwise. Thus,

$$EV(27) = \frac{1}{38}(35) + \frac{37}{38}(-1) = \frac{-1}{19} = -.0526.$$

Hence we see that for every dollar bet on a single number (as 27 is no different than any other single number), we expect to lose a little over 5 cents! Another way to interpret this is that the House has a slightly better than 5% advantage on this kind of bet. □

Example 2.3.2. *What is the expected value of a $1 bet on red?*

Solution: This time $P(\text{red}) = \frac{18}{38}$ and the House odds are 1:1. Hence, our expected value computation is

$$EV(\text{red}) = \frac{18}{38}(1) + \frac{20}{38}(-1) = \frac{-1}{19} = -.0526.$$

The expected value is exactly the same as for the single number bet! □

TABLE 2.3:   True and House Odds in American Roulette

Type of Bet	True Odds	House Odds
Red (or black)	20:18	1:1
Even (or odd)	20:18	1:1
1-18 (or 19-36)	20:18	1:1
Any 12 no. (column or dozens)	26:12	2:1
Any two rows	32:6	5:1
Any four no. square	34:4	8:1
Any row	35:3	11:1
Two adjacent no.	36:2	17:1
Single no.	37:1	35:1

Example 2.3.3. *What is the expected value of the combination of a $1 bet on number 4 and a $1 bet on black?*

Solution: Here we must take two bets into consideration. The single number 4 is also black, so if 4 is the number then we will win both bets. But we can also win the black bet with 17 other numbers. Hence, with probability 1/38 we will win both bets, with probability 17/38 we will win black, but not the number 4, and in all other cases we lose both bets. Now our expected value computation becomes

$$EV(\text{both bets }) = \frac{1}{38}(35+1) + \frac{17}{38}(1-1) + \frac{20}{38}(-2) = \frac{-4}{38} = \frac{-2}{19}.$$

Thus we see that the combination of two bets at $1 each produces an expected value that is twice what a $1 bet produced. In other words, our per dollar expected value is unchanged! □

We will consider some of the other bets and combinations of bets in the exercises. However, we shall see that the per dollar expected value for those bets will be the same. Thus, in playing American Roulette, the House will always hold a better than 5% advantage over the players, no

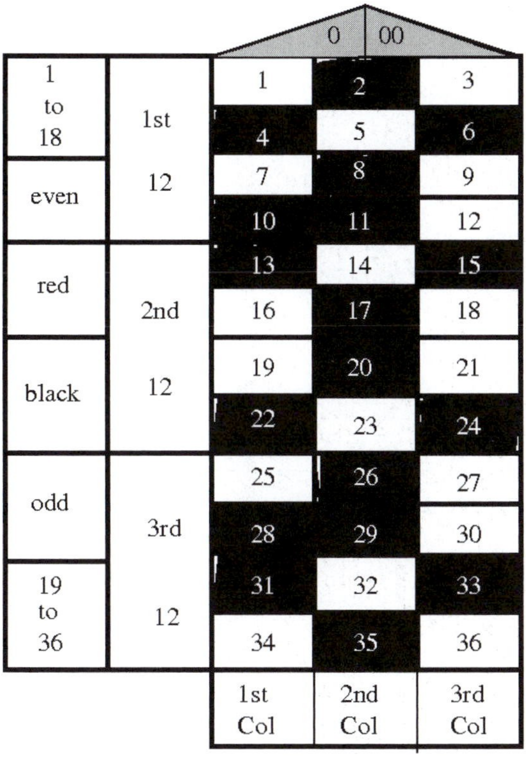

FIGURE 2.2: Roulette table.

matter how they bet. This says that, with long-term play, we should expect to lose an average of about 5 cents for every dollar we bet! This prospect makes roulette a fairly unattractive game to many people.

Exercises:

2.3.1 What is the probability of 0 or 00 in roulette?

2.3.2 A European Roulette wheel has 37 numbers (no 00). What is the expected value of a $1 bet on number 27 in European Roulette?

2.3.3 What is the expected value of a $1 bet on red and a $1 bet on number 3 in European Roulette?

2.3.4 What is the expected value of a $1 bet on red and a $1 bet on number 2 in European Roulette? What is the expected value of the same bet in American Roulette?

2.3.5 What is the expected value of a $1 bet on a single column in American Roulette?

2.3.6 Use Table 2.3 to verify the true odds of a bet on any two rows? On any four number square? On any 12-number combination?

2.3.7 What is the expected value of a $1 bet on any row?

2.3.8 What is the expected value of a $1 bet on two adjacent numbers?

2.4   Craps

This is often thought of as the ultimate dice game. *Craps* remains a popular casino game, as it has been for many years. Part of its popularity stems from the wide range of possible bets. It also offers fast action, thrills and lots of noise and excitement.

The game can be traced to many old versions but probably owes its modern popularity to the widespread play it received among soldiers during World War II (see [16]).

The basics of the game are simple. One player (*the shooter*) rolls a pair of dice. All players bet on the outcomes of these rolls. On the first roll the shooter is a winner provided he or she rolls a total of 7 or 11 (called a *natural*). (Recall, we showed in Example 1.8.1 that this happens with probability 8/36.) The shooter is a loser on the first roll if he or she rolls a 2, 3 or 12 (called craps). This happens with probability 4/36 as you may easily verify from Table 1.2. If the shooter fails to roll a natural or craps, say they roll a 9 instead, then the number rolled becomes the so-called *point*. The point is the number rolled on the initial roll other than 7, 11, 2, 3 or 12. The shooter then begins on a series of rolls to either roll the point again and win (or *pass*) or roll a 7 and lose (that is *don't pass*). Any other roll is ignored as far as ending the game is concerned. This is significant as it changes our sample space! The shooter continues to roll until one of those two outcomes occurs.

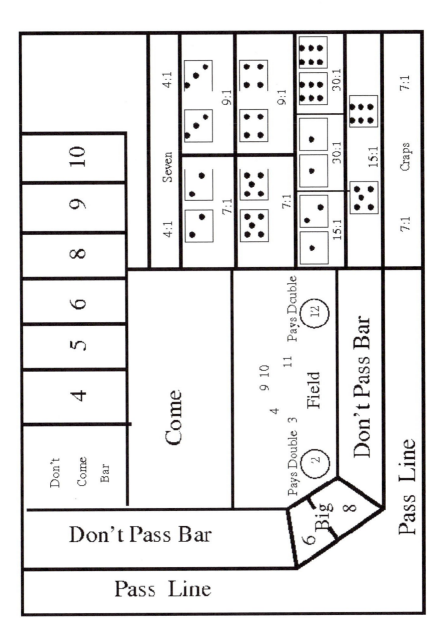

FIGURE 2.3: A typical craps table.

2.4.1   Street Craps

A reasonable starting place is with the game that was popular with the soldiers. This is often called *street craps*. In this game there are two fundamental bets, that the shooter wins, called *betting the pass line*, or that the shooter loses, called *betting the don't pass line*. Many other bets are possible, but since there is no House, someone else must be willing to take the bet. Betting on either the pass line or the don't pass line is an even money bet. Usually the shooter is willing to take all bets against himself/herself. We begin our analysis here.

TABLE 2.4:   Partial Table for Events and Probabilities for Craps

1st Roll	P(1st roll)	P(1st then pass)	P(1st then don't pass)
7 or 11	$\frac{8}{36}$	$\frac{8}{36} = .2222$	.0000
$2, 3, 12$	$\frac{4}{36}$	.0000	$\frac{4}{36} = .1111$
4	$\frac{3}{36}$	$\frac{3}{36}\frac{3}{9} = .0278$	$\frac{3}{36}\frac{6}{9} = .0556$
5	$\frac{4}{36}$	$\frac{4}{36}\frac{4}{10} = .0444$	$\frac{4}{36}\frac{6}{10} = .0667$
6			
8			
9			
10			

As there are several ways for the shooter to win or lose, the analysis must consider all these cases. The easy case is if the shooter wins or loses on the first roll. We have already determined these probabilities and they are shown in Table 2.4. The more interesting parts of play come when the shooter establishes some point and must continue to roll.

Example 2.4.1. *Suppose the shooter has established a point of five. What is the probability the shooter will win, that is, roll another 5*

before rolling a 7?

Solution: In our problem the game has simplified. All that matters now is rolling a 5 or a 7. So our sample space has also changed. It consists of the 6 ways to roll a 7 and the 4 ways to roll a 5. Since each outcome is equally likely, we see that

$$P(\text{shooter makes the point 5}) = \frac{4}{10}. \quad \square$$

Example 2.4.2. *What is the probability the point becomes 5 and the shooter also succeeds in making the point?*

Solution: Now we have dependent events. We must first know the probability the point will be 5. This is easily seen to be 4/36. Combining this with the result of our last problem and using Rule 1.4.6 (3) we see that

$$P(\text{point 5 AND pass}) = P(\text{point becomes 5})P(\text{pass} \mid \text{point is 5})$$

$$= \frac{4}{36} \times \frac{4}{10}$$

$$= .0444. \quad \square$$

In Table 2.4 we also show the results for making a point of 4. Similar computations can be done for each of the other possible points (see exercises).

Problem 2.4.1. *Complete Table 2.4.*

Once you have completed Table 2.4 you will be able to verify that $P(\text{shooter passes}) = .4928$ while $P(\text{don't pass}) = .5071$. Thus, in street craps there is a distinct advantage in betting against the shooter! If the shooter is taking your bet against him or her, this can potentially cause tensions to build.

2.4.2 Casino Craps

When the game of craps became a casino game, there was one immediate problem that had to be fixed. Betting the don't pass line had a probability of more than .5 and so it would have a positive expected value. No casino will allow such an advantage to the player, as a smart player would make no other bet!

Example 2.4.3. *How do casinos negate this advantage in betting the don't pass line?*

Solution: The casinos found a fairly easy way to change the betting advantage. One common solution is to change the losing payoffs on the initial roll. It is common for the casino to change an initial roll of double 6 from a losing roll to a *push*, that is, neither a win nor a loss in terms of money bet. The shooter loses and all those betting the pass line also lose, but those betting the don't pass line receive no payments (but also do not lose their bets). This decreases the probability of winning on the don't pass line by $1/36 = .0278$.   □

Example 2.4.4. *How does the expected value of betting the no pass line in street craps change under the rules for casino craps?*

Solution: We can compute the expected value (per dollar) for the pass line bet E_1 and the don't pass bet E_2 in street craps. The pass line expected value is

$$EV(E_1) = .4928(1) + .5071(-1) = -.0143$$

while the don't pass expected value is

$$EV(E_2) = .4928(-1) + .5071(1) = .0143.$$

But under the changes for casino craps, the positive expected value of .0143 would be decreased by .0278 giving

$$EV(\text{casino don't pass}) = .0143 - .0278 = -.0135.$$

Thus, under this change the two expected values are nearly equal and there is certainly no major advantage in betting against the shooter. More importantly for the casino, the expected value is now an acceptable negative amount.   □

2.4.3   Other Bets

As we stated earlier, craps offers a vast array of bets, with a vast array of odds. Some of these bets are better than others as we shall see. One of the best bets in craps is the *free odds* bet.

If on the initial roll the shooter establishes a point, then a bettor may back up a pass line bet (or no pass bet) with up to an equal additional amount bet at the true mathematical odds for making the point, not the casino house odds.

Suppose the bettor has bet $20 and the established point is 5. Then the bettor may now place an additional $20 on free odds (betting that 5 will appear before 7) at the true odds of 3:2. This additional bet has an effect on the expected value, helping move the expected value per dollar bet closer to zero and thus, closer to a fair bet (provided you always take free odds — which is definitely good advice). If the point of 5 wins, the bettor will now win $20 from the initial bet and $30 from the free odds bet! However, rolling a 7 will cause the bettor to lose $40.

Example 2.4.5. *What may a bettor win or lose if he or she bet $10 on the pass line, 6 is the established point, and the bettor backs the pass line bet with a $10 free odds bet?*

Solution: Of course the bettor may lose a total of $20, as that is what the bettor has wagered. As to the winnings, they can win $10 on the pass line as it always pays 1:1. The free odds bet when 6 is the point has true odds of 6:5, as there are six ways to roll a 7 and five ways to roll a 6. Thus, the $10 bet on free odds would return $12. Hence, our bettor can win a total of $22. □

Example 2.4.6. *We now wish to do a general computation of our expected value, assuming a standard $10 bet on the pass line, followed up with a $10 free odds bet, no matter what the point happens to be.*

Solution: We let the random variable $X(point)$ be the value we win with a given point. Using the symmetry of the point probabilities of 4 and 10, 5 and 9, and 6 and 8, we can simplify our computation to the following:

$$
\begin{aligned}
EV(X) = \ & .2222(10) + .1111(-10) \quad \text{natural and craps} \\
& + .0556(30) + .1111(-20) \quad \text{4 and 10} \\
& + .0889(25) + .1334(-20) \quad \text{5 and 9} \\
& + .1262(22) + .1514(-20) \quad \text{6 and 8} \\
= \ & -.1414
\end{aligned}
$$

We must keep in mind that we are betting more using free odds, so in order to compare expected values, we should do it on a per dollar basis. In the exercises you will be asked to show that the expected amount wagered under this strategy is \$16.67 (see Exercise 2.4.9). Thus, this strategy has an expected per dollar wagered value of

$$\frac{-.1414}{16.67} = -.0085.$$

That is,

$$\frac{-.85 \text{ cents}}{\text{dollar}}.$$

This is a small amount per dollar and is by far the closest to a fair bet (zero expected value) you can find in any casino game. □

TABLE 2.5: Odds and Casino Edge for Some Other Craps Bets

Bet	Prob. Win	House Odds	House Edge
Place	varies	varies	
Field	.444	varies	5.6%
Big 6/Big 8	.455	1:1	9.1%
Hard 6/Hard 8	.091	9:1	9.1%
Hard 4/Hard 10	.111	7:1	11.1%
Craps	.111	7:1	11.1%
Hard 2/Hard 12	.028	29:1	16.7%

Table 2.5 shows some of the other bets possible in craps and the house advantages. Note that several of these are very large advantages. We shall briefly explain these bets.

The *Field* bet is that on the next roll of the dice, a 2, 3, 4, 9, 10, 11 or 12 will be rolled. All other numbers lose. A *place* bet is a bet on one of the possible point numbers. To win, your number must be rolled before a 7 is rolled. Place bets are paid as follows: a 4 or 10 gets paid at 9:5, while a 5 or 9 is paid at 7:5, and finally 6 or 8 is paid at 7:6. The *Big 6 or Big 8* bet is that a 6 or 8 will be rolled prior to a 7. Thus, it is like a place bet only paid at the lower rate of 1:1. Hence, it is a very bad bet. The craps bet is a bet that the shooter will roll a 2, 3 or 12 on the next roll. It is paid at 7:1. The 12 bet is a one roll bet that the shooter will roll a 12. It is paid at 30:1. The *hard bets* are bets that a double will be rolled. Depending on the number you select, you are betting that the doubles for that come up before any other combination that yields that particular sum, or before a 7 is rolled. The odds are 7:1 for hard 4 (double 2s) and hard 10. The house odds are 9:1 for a hard 6 or hard 8.

The exercises will consider these bets in several ways.

Exercises:

2.4.1 Determine the true odds of making the point when the point is 6, 8, 9, or 10.

2.4.2 Use the previous problem to complete Table 2.4.

2.4.3 Compute the true odds of making a field bet and compare this to the house odds. Determine the house advantage for a field bet.

2.4.4 Determine the true odds for the various place bets.

2.4.5 Use the last problem to determine the house advantage on a place bet for each of the possible points.

2.4.6 Determine the true odds of a hard 6.

2.4.7 Determine the true odds of a hard 8 and a hard 10.

2.4.8 Use the previous two exercises to help in determining the house advantage on a hard 8 bet, a hard 10 bet and a hard 12 bet.

2.4.9 Compute the expected amount bet if you play the pass line betting $10 and always take free odds for $10 when possible.

2.4.10 Consider a casino in which the push bet on craps is a total of 3. Compute the house advantage on a don't pass bet in this casino.

2.4.11 Verify the house advantage (see Table 2.5) for a field bet.

2.4.12 Verify the house advantage for a Big 6 bet.

2.4.13 Verify the house advantage for a Hard 10 bet.

2.5   Let's Make A Deal — The Monty Hall Problem

The *Monty Hall Problem*, as it is now known, derived directly from a TV game show called *Let's Make A Deal* (see [26]) whose host was Monty Hall. Reruns of the show can still be seen on several networks. In each episode the final challenge given to a contestant was the following.

Example 2.5.1 (The Monty Hall Problem). *A contestant is given the choice of three doors: behind one door is a car (or some other great prize); behind the other two doors, goats (or other terrible prizes). The contestant is allowed to pick a door, say door No. 1, and Monty Hall, who knows what's behind the doors, opens another door, say door No. 3, which has a goat behind it. He then asks the contestant, "Do you want to pick door No. 2 or stay with your original choice?"*

Question: Is it to your advantage to switch your choice?

This question gained national attention when a widely known statement of the problem appeared in a letter to Marilyn vos Savant's *Ask Marilyn* column in *Parade* magazine [27].

Most people assumed that with two doors remaining unopened, there was a .5 probability of the car being behind either door and hence no real reason to switch doors. However, vos Savant argued differently in her column. This caused a significant stir and many letters of complaint, including some from mathematicians. However, vos Savant's argument has been justified now by many others. It is a matter of conditional probability. For more on the controversy see [26] and [27].

An intuitive argument why you should switch doors is the following: the probability you initially select the door with the car behind it is certainly 1/3. Thus, there is clearly a probability of 2/3 that the car is behind one of the doors you did not select. You know a goat is also behind one of the doors you did not select. Monty Hall showing you the goat you know is there does nothing to change the original probability you selected correctly. Thus, there is still only a 1/3 probability your door is correct and a 2/3 probability you were incorrect, and hence a probability of 2/3 that the car is behind the other door. Thus, to take advantage of the probabilities you should switch doors.

Note, the website at http://math.ucsd.edu/ crypto/Monty/monty.html has an online simulation of the game which you can play. It also gives statistics for all others that have played, thus you can see what the best move has been during play.

Solution: We now formally solve the Monty Hall Problem using conditional probabilities. Let C_i be the event that the car is behind door i, for $i = 1, 2, 3$. Let H_{ij} be the event that you select door i and Monty Hall opens door j.

Clearly, $P(C_i) = 1/3$, for $i = 1, 2, 3$. Now, consider when Monty Hall opens a door with a goat behind it. If two such doors are available, they are equally likely to be selected, while if only one door has a goat behind it, Monty Hall must show you that door. This determines the conditional probability of H_{ij}. Assuming the car is behind door C_k, then

$$P(H_{ij} \mid C_k) = \begin{cases} 0 & \text{if } i = j \text{ it's your door already} \\ 0 & \text{if } j = k \text{ he won't show the car} \\ 1/2 & \text{if } i = k \text{ either door, it's random} \\ 1 & \text{if } i \neq k \text{ and } j \neq k \text{ only 1 door to open.} \end{cases}$$

Assume, without loss of generality, you select door 1 and Monty Hall opens door 3. Now, from above we see that $P(H_{13} \mid C_1)) = \frac{1}{2}$ and $P(C_1) = \frac{1}{3}$. Then

$$P(C_1 \mid H_{13}) = \frac{P(H_{13} \mid C_1)P(C_1)}{P(H_{13})}$$

$$= \frac{\left(\frac{1}{2}\right)\left(\frac{1}{3}\right)}{P(H_{13})}$$

$$= \frac{\frac{1}{6}}{P(H_{13})}.$$

But, now we know by the Total Probability Formula that

$$P(H_{13}) = P(H_{13} \mid C_1)P(C_1) + P(H_{13} \mid C_2)P(C_2) + P(H_{13} \mid C_3)P(C_3)$$

$$= (1/2)(1/3) + (1)(1/3) + (0)(1/3)$$

$$= 1/2.$$

Thus,

$$P(C_1 \mid H_{13}) = \frac{1/6}{1/2} = \frac{1}{3}.$$

That is, the probability the car is behind door 1 when you have selected door 1 and Monty Hall has opened door 3 remains at 1/3, just as it was before door 3 was opened. Showing you the goat behind door 3 changed nothing!

Now, the probability of winning by switching to door 2 is $P(C_2 \mid H_{13})$. We can obtain this value from

$$1 = P(C_1 \mid H_{13}) + P(C_2 \mid H_{13}) + P(C_3 \mid H_{13}).$$

Since $P(C_3 \mid H_{13}) = 0$ (Monty does not show the car!), therefore

$$P(C_2 \mid H_{13}) = 1 - P(C_1 \mid H_{13}) = 1 - 1/3 = 2/3.$$

That is, the probability the car is behind door 2 given you selected door 1 and Monty Hall opened door 3 is 2/3. Thus, the probability of winning the car by switching is 2/3! The argument remains the same no matter what door the contestant selects or what door Monty Hall opens.   □

Of course we saw a simple-minded argument that the probability of winning the car when you selected initially was 1/3 and there was a probability of 2/3 that the car was behind one of the other two doors, and nothing has really changed since then. So with probability 2/3 you should switch to the available door. The intuitive argument has now been shown to be correct. Our formal argument using conditional probabilities has confirmed our reasoning.

Exercises:

2.5.1 Create a table of the possible outcomes of the Monty Hall problem to confirm that there is a 2/3 probability of winning by switching doors.

2.5.2 How does the problem change if Monty Hall does not know where the car is? Create a table as in the previous problem to demonstrate this change.

2.5.3 Stacey and two other contestants on a game show must answer questions. They must answer three questions from a card drawn at random from a set of 20 cards. There are 8 favorable cards for Stacey (she knows the answer to all three questions on these cards). Stacey wins a major prize if she can answer all three questions correctly. What is the probability Stacey wins the prize if:

1. Stacey draws first.

2. Stacey draws second (no replacement of cards).

3. Stacey draws third?

2.5.4 There are two bowls. The first bowl contains three tickets to the superbowl and four tickets to a square dance. The second bowl contains six tickets to the superbowl and three tickets to the square dance. At random, you take one ticket from bowl one and place it in bowl two. Now you select a ticket from bowl two. What is the probability the ticket you selected from bowl two is a superbowl ticket?

2.5.5* How does the Monty Hall Problem change if we use four doors instead of three? That is, you have a choice of two doors to switch to instead of just one.

2.6 Carnival Games

In this section we will take a brief look at a couple of other games commonly found at carnivals, Chuck-a-Luck and Poker Dice.

The game of Chuck-a-Luck is very simple and I remember seeing it at carnivals and fairs that I attended as a child. The numbers 1, 2,..., 6 appear in individual squares. You may bet on any (or all) of them. Once bets are placed, three dice are rolled (or as I remember them, flipped in a bird cage type device). The payoff depends on the roll. If your number fails to appear on any die you lose your bet. If your number appears on one die you will be paid at 1:1. If your number appears on two dice, you will be paid at 2:1 and if your number appears on all three of the dice you will be paid at 3:1.

Example 2.6.1. *You are playing Chuck-a-Luck. You bet $1 on the number 5. The three dice are rolled and the outcome is 3, 5, 5. What are you paid?*

Solution: In this case, as you bet the number 5, you are paid at a rate of 2:1. Thus, you receive your original bet back and you also receive $2 in winnings. □

Example 2.6.2. *Compute the number of winning outcomes for a bet in Chuck-a-Luck.*

Solution: We already know that in rolling three dice there are $6^3 = 216$ possible outcomes. Thus, we must determine the number of these that match one, two or three times with our selected number.

For three matches, all three dice must equal our selected number and this can happen in only one way. For two matches in three dice we must first choose the dice that will match. This can be done in $\binom{3}{2}$ ways. Now the remaining die can have any of the other five numbers. Thus, there are $\binom{3}{2} \times 5 = 15$ outcomes with exactly two matches. Finally, for exactly one match there are $\binom{3}{1}$ ways to select the die that will match. Each of the other two dice can be any of the other five numbers. Thus, for one match there are $\binom{3}{1} \times 5^2 = 75$ possible outcomes. $\square$

The interesting question here is the expected value of a \$1 bet at Chuck-a-Luck (see Exercise 2.6.2). Counting the winning rolls is the key to answering this question.

The second carnival game is Poker Dice. The name Poker Dice comes from the tool in use, an ordinary die with the symbols 9, 10, J, Q, K, A on the six sides, rather than the normal integers. The bettor chooses two different faces from among the six choices. Then five poker dice are rolled. The bettor wins as follows:

If both chosen faces appear, the bettor is paid at a rate of 1:1. Otherwise, the bettor loses.

Example 2.6.3. *What is the expected value of a \$1 bet that an ace and king will appear in the game of Poker Dice?*

Solution: To solve this problem we recognize three possible outcomes for the roll of five dice.

1. No ace or king appears on the five dice.

2. One of the ace or king appears, but not the other.

3. Both an ace and king appear.

It is fairly easy to determine the probability of (1) happening. There are four ways for no ace or king to happen on any one die and so the probability is:

$$(4/6)^5 = (2/3)^5 = .1317.$$

In order to determine the probability of one of the ace or king appearing, we first suppose no king appears (event E_1), but at least one ace appears (event E_2). We seek $P(E_1 \cap E_2)$. Now,

$$P(E_1) = (5/6)^5 = .40188.$$

In those outcomes with no king at all, the other five possibilities all occur with equal probability. Thus, the chance no ace appears is $(4/5)^5$. Thus, the probability of at least one ace given that no king appears is

$$P(E_2 \mid E_1) = 1 - (4/5)^5.$$

Now, $P(E_1 \cap E_2) = P(E_1)P(E_2 \mid E_1)$. Thus,

$$P(E_1 \cap E_2) = (.40188)(.67232) = .2702.$$

The computation for no ace but at least one king is clearly the same. Thus, our desired probability is

$$2(.2702) = .5404.$$

Now, the easiest way to compute the probability of Case (3) happening is using the fact the three probabilities sum to one. That is,

$$1 = .1317 + .5404 + P((3) \text{ happens}).$$

Thus, $P((3) \text{ happens}) = 1 - .1317 - .5404 = .3279$. As this is just less than $1/3$, and we are paying at even money, the carnival should be happy there is a positive expected value for this game! □

Exercises:

2.6.1 Determine the probabilities for one match, two matches and three matches in Chuck-a-Luck.

2.6.2 Find the expected value of a $1 Chuck-a-Luck bet. (Hint: see Exercise 1.4.7).

2.6.3 In Chuck-a-Luck, determine payouts X, Y and Z for one match, two matches, and three matches, respectively, that would make it a fair game.

2.6.4 What is the expected value of a $1 bet in Poker Dice?

2.6.5 What is the expected value of a $1 bet at Poker Dice if we instead use six dice?

2.6.6 What is the expected value of a $1 bet at Poker Dice if we use seven dice?

2.6.7 What payout for poker dice would make it a fair game?

2.7   Other Casino Games

In this section we take brief looks at some other casino games, and some questions of interest to these games. We try to consider some new questions in order to broaden the impact of our tools.

2.7.1   Caribbean Stud Poker

Unlike most poker games, Caribbean Stud Poker is a game of luck, with no real skill involved in its play. This is because the rules are set and your only decision is to fold or bet a fixed amount. Essentially, each player plays against the dealer. Before you see your cards, you make an initial bet (the *ante*). Then comes the deal of five cards to each player and the dealer, with the dealer's last card face up. Now when it is your turn, you fold if you do not believe your hand can win, or you *raise* by exactly twice the ante, if you believe your hand has a chance to win. This completes the betting.

At this point you really have two separate bets, the ante and the raise. The dealer then shows his hand. If the dealers hand is *nonqualifying*, that is, worse than an AKxxx (that is ace, king and 3 small nonmatching cards), you win no matter what you hold, but only even money on the ante. The raise is a push. This in fact happens often (about 44% of the time).

When you have raised and the dealer holds a qualifying hand of AK or better, then both your bets are in play. Now, if the dealers hand is better, you will lose both bets. If your hand is better, you are paid even money on the ante and you are paid on the raise at varying rates based on the quality of your hand (see Table 2.6).

By now we know that all casino games favor the house. But there are more questions that can be asked here. To play this game, you would

TABLE 2.6: Pay Table for
Raises in Caribbean Stud Poker

Royal flush	100 to 1
Straight flush	50 to 1
Four of a kind	20 to 1
Full house	7 to 1
Flush	5 to 1
Straight	4 to 1
Three of a kind	3 to 1
Two pair	2 to 1
All other	1 to 1

like to know things like:

- Which hands should I raise?

- How big is the house advantage?

- Can I learn anything from the one exposed card in the dealer's hand?

These are not simple questions because the cards you hold do have an effect on the cards the dealer holds. If you hold three aces and two kings, the dealer's chances of a qualifying hand have been decreased! Also, because my hand and the dealer's hand are not independent, the exact computations are rather involved.

We will concentrate on the first question and use approximations rather than exact computations in order to make the job more reasonable.

The question of whether to raise or fold is one of comparison of the expected values for the possible choices. This is a bit complicated in that we need to count the number of dealer hands that are non-qualifying, qualifying but worse than our own hand, qualifying but better than our own hand, and qualifying but tied with our hand.

The total for all these possibilities is $N = \binom{47}{5}$, as this is the number of possible hands the dealer can have once our own five cards are removed. Let's make a few simple assumptions to help in our counting. Let w be the number of dealer hands worse than AK. Let q be the number of dealer hands of AK or one pair, that are still worse than our hand. Finally, let t be the number of dealer hands that push against our hand. For simplicity, we assume an ante of 1 (think of it as one unit bet, although in most casinos the minimum bet is \$5), so our raise will be 2 (units). Now, we can do the expected value computation as follows:

$$EV(\text{raise}) = \frac{w}{N}(1) + \frac{q}{N}(3) + \frac{t}{N}(0) + \left(\frac{N - w - q - t}{N}\right)(-3).$$

Thus,

$$EV(\text{raise}) = \frac{4w + 6q + 3t}{N} - 3.$$

Then, clearly we should raise when this exceeds $EV(fold) = -1$. Thus, raise when

$$\frac{4w + 6q + 3t}{N} - 3 > -1,$$

or

$$4w + 6q + 3t > 2N. \tag{2.1}$$

We can do these counts for particular hands to determine if they satisfy Equation (2.1). Some of these types of questions are included in the exercises.

2.7.2   Keno

Keno is a lottery type game in which a player marks several numbers from the numbers 1 to 80. Then, the casino randomly selects 20 numbers. If many of the players' numbers are selected relative to the number initially marked (we say catching those numbers), then the player wins. The more selected numbers appearing among the 20 numbers, the more the player wins. Some typical bets and payoffs are shown in Table 2.7.

Example 2.7.1. *What is the probability of catching five numbers when you bet eight numbers?*

TABLE 2.7: Some Keno Bets and Payoffs

	Play 7	Play 8	Play 9
Catch 4	$2	$0	$0
Catch 5	$32	$12	$4
Catch 6	$750	$210	$50
Catch 7	$20,500	$3,100	$900
Catch 8		$50,000	$12,500
Catch 9			$50,000

Solution: To answer this we begin by looking at the 80 numbers as being partitioned into two parts, the 20 good numbers and the 60 bad numbers. If the player is to catch five out of eight, then the five must be from the good numbers and the other three from the bad numbers. There are $\binom{20}{5}$ ways of selecting the five good numbers and there are $\binom{60}{3}$ ways of selecting the three bad numbers. Thus, there are

$$\binom{20}{5}\binom{60}{3} = 530,546,880$$

ways of selecting the eight numbers for the player so that five will match.

Now, the total number of ways of selecting eight numbers is

$$\binom{80}{8} = 28,987,537,150.$$

Thus,

$$P(\text{Catch 5}) = \frac{530,546,880}{28,987,537,150} = .0183.$$

□

Example 2.7.2. *Compute the probabilities of catching 6, 7 or 8 numbers when you play 8 numbers.*

Solution: Following what we did in the previous problem, we see that

$$P(\text{Catch 6}) = \frac{\binom{20}{6}\binom{60}{2}}{\binom{80}{8}} = .00237,$$

$$P(\text{Catch 7}) = \frac{\binom{20}{7}\binom{60}{1}}{\binom{80}{8}} = .000160,$$

and

$$P(\text{Catch 8}) = \frac{\binom{20}{8}\binom{60}{0}}{\binom{80}{8}} = .00000435.$$

$\square$

Example 2.7.3. *What is the expected value of a $2 bet on 8 numbers in Keno?*

Solution: Here we use the information from the previous two problems and Table 2.7, so that

$$
\begin{aligned}
EV(8) = \ & 10(.0183) + 208(.00237) + 3098(.000160) \\
& + 49998(.00000435) + (-2)(.979) \\
= \ & -0.79.
\end{aligned}
$$

But this was based on a $2 bet. Thus, we see that the per dollar expected value of Keno 8 bet is $-.399$. This is clearly a huge disadvantage!

2.7.3   Blackjack

The game of blackjack is a very popular one at most casinos in the United States. The concept of the game is simple. Each player plays against the dealer. The object is to get a count as close to 21 as possible, without exceeding that number. The count is the face value of the card, with aces counting 1 or 11, and face cards (J, Q, K) counting 10. Each player and the dealer are dealt two cards, with the dealer's second card face up. Now, in turn, each player may stand (play just the cards they have) or take additional cards. After each new card that does not cause the player to exceed 21, they again have the choice of standing or taking another card. After each player has had his opportunity, the dealer turns over her other card. If the dealer's sum is below 17, the dealer must take an additional card. Once the dealer's sum is in the 17 to 21 range, the dealer must stop taking additional cards. Players whose totals do not exceed 21, but do exceed that of the dealer are paid at 1:1. If a player is dealt a blackjack, namely a two card hand totaling 21, he is paid at 3:2. Players that exceeded 21 lose immediately, even if the dealer later also exceeds 21. There are more rules, but these are sufficient for what we shall do in this section.

Some versions of blackjack come with a side bet (usually $1), in which the player receives a bonus payment if his hand is of a very particular type. Two bonus types are common, *The Royal Hand* being a suited king and queen combination, or the *hard 21*, being three 7s.

Example 2.7.4. *What is the probability of being dealt The Royal Hand?*

Solution: There are eight kings and queens in the deck. So the first part of our computation is the probability of getting one of these cards on the first card, that is $8/52$. Now, on the second card, only the match to the pair is possible. This clearly has probability $1/51$. Now by the Multiplication Principle,

$$P(\text{Royal Pair}) = (8/52) \times (1/51) = .00302.$$

$\square$

Problem 2.7.1. *What is the probability of being dealt the hard 21?*

Solution: See Exercise 2.7.11. $\square$

A second common option in blackjack is called *insurance*. This option is offered to the players when the dealer has an ace showing. The player may bet up to one half his initial bet as the insurance bet. The insurance bet is paid at 2:1 when the dealer does have a blackjack.

Example 2.7.5. *The dealer has an ace showing and suppose you are holding a king and a jack. Assuming a single deck is in use, what is the expected value of an insurance bet of $1?*

Solution: There are 16 cards that can complete the dealer's blackjack, but we hold two of them. In this case the probability the dealer has a blackjack is

$$P(\text{dealer blackjack}) = P(\text{ins. wins}) = 14/49 = 2/7.$$

Further,

$$P(\text{no blackjack}) = P(\text{ins. loses}) = 1 - 2/7 = 5/7.$$

Thus,

$$EV(\text{ins.}) = 2(2/7) + (-1)(5/7) = -.143. \quad \square$$

Example 2.7.6. *Again the dealer has an ace showing. This time you hold a 3 and 8 and your friend at the table holds a 4 and 5. Now what is the expected value of a $1 insurance bet?*

Solution: Given the information available we see that $P(\text{ins wins}) = 16/47$ and $P(\text{ins loses}) = 31/47$. Thus,

$$EV(\text{ins.}) = (2)(16/47) + (-1)(31/47) = .021.$$

Hence, it is possible to have a positive expected value for the insurance bet. $\quad \square$

Exercises:

2.7.1 At Caribbean Stud Poker, you hold $AK876$. How many hands without a pair will defeat your hand?

2.7.2 In Caribbean Stud Poker, you hold the minimum qualifying hand of $AK432$. Compute the expected value of raising with this hand. Compare this to the expected value of folding (use 1 unit ante and 2 units for raise).

2.7.3 In Caribbean Stud Poker you hold $AKQJ9$, the top $AKxxx$ hand. Compute the expected value of a raise. (Hint: here you may assume 49.9% of all hands are one pair or better.)

2.7.4 Compute the expected value of a 7 number bet at Keno.

2.7.5 What is the probability of catching 6 numbers in a 9 number bet at Keno?

2.7.6 What is the probability of catching 7 numbers in a 9 number bet at Keno?

2.7.7 At blackjack, what is the probability you are dealt a blackjack (21 in two cards)?

2.7.8 At blackjack, what is the probability you are dealt 20 in two cards?

2.7.9 At blackjack, the dealer has an ace showing, you hold 10, 7, while the other three players hold 5, 8 and J, 4 and 3, 3. Again assuming single-deck play, what is the expected value of a $1 insurance bet?

2.7.10 In blackjack, what payoff would make The Royal Hand $1 bet a fair bet?

2.7.11 In blackjack, what is the probability of a hard 21 and what payoff would make the hard 21 bet of $1 a fair bet?

2.7.12 In blackjack, a *surrender* bet is one where the player, after receiving the initial two cards, withdraws half his bet, surrendering the other half of the bet to the house without taking any cards. Compute the expected value for surrendering a $2 bet versus the expected value of not surrendering? (Hint: assume a probability of p for winning the hand if you play.)

2.7.13* At Caribbean Stud Poker, you hold $JJ654$. How many hands with one pair will defeat your hand?

2.7.14* In Caribbean Stud Poker, you hold $AAKKQ$. What is the probability the dealer has a qualifying hand?

2.8 Backgammon

The game of *backgammon* traces its roots back to 3000 BC where Mesopotamians played a board game with four-sided pyramidal dice. The game was certainly popular in medieval Europe and has remained a game that combines skill and the luck that comes with dice games [36].

We will not attempt to completely describe the rules. There are many outstanding books about backgammon and the interested reader might wish to consult one (see for example [22] or [42]). Instead, we will explain only as much as necessary to allow us to study certain aspects of the game.

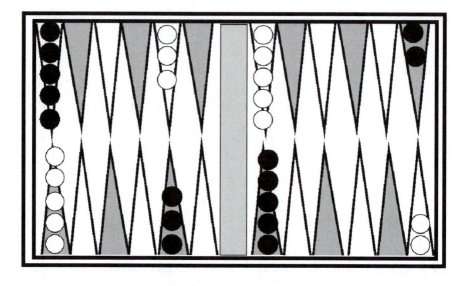

FIGURE 2.4:   Backgammon board in opening position.

The object of the game is to move all your pieces (you are either white or black; see Figure 2.4) around the board and eventually off the board. First you must get all your pieces into an area called your *inner table*, the final six-pointed region. In Figure 2.4 this is the set of "points" at the bottom right for black and upper right for white. You win by "bearing off" your pieces from your inner table into your *home* (off the board) before your opponent does. So in this sense the game is a race. Opponents move in opposite directions around the board, which creates instances for conflict. Thus, backgammon offers much in the way of strategy and action.

Moves in backgammon are made based on the roll of two dice. You may move any one of your pieces exactly as far (counting points on the board) as the spots shown on one die. Then repeat the process for the other die. If you roll doubles you may make double the moves on each die, or equivalently, four times you may move a piece the amount shown on one die. In addition to moving, you may hit an opponent if one of your pieces lands (ends its move) on a space occupied by a single piece belonging to your opponent, called a *blot*. When this occurs, your opponent's piece is removed to the *bar*, which is the center area of the board between the points. This piece essentially must start over, moving into your opponent's inner table and then moving fully around the board. A board position, or *point*, becomes safe when two or more (there is no limit on how many) of a player's pieces occupy the point. At that time no opponent's pieces may land on that point. Note that if your only move would land you on an opponent's safe point, then you cannot move and your turn ends. Games of backgammon are played for an initial stake, which can be changed in several ways, one of which is through the use of a *doubling cube*. We will discuss the doubling cube in more detail later. Our primary focus will be on aspects of the game we can analyze using elementary probability theory and expected values.

2.8.1 Hitting Blots

Perhaps the easiest aspect of the game to consider is the act of hitting blots. The work we did earlier on dice pairs will be helpful here (see Section 1.2), but we must also keep in mind the special nature of rolling doubles in backgammon when we do probability computations.

The first thing that we notice is that our ability to hit a blot depends on where that blot is with respect to our own pieces. For example, if a blot is one point away, then there are 11 possible rolls in which a 1 occurs on at least one die, and hence the probability of being able to hit that blot is $11/36$. But if the blot is 12 points away, then our only hope of hitting it is by rolling double 6, or double 3 or double 4. Thus, the probability of hitting that blot is $3/36$. There is a considerable difference in these values. To truly understand what is happening here, the reader needs to complete Table 2.8 (see the exercises).

2.8.2 Off the Bar

Being forced onto the bar can be a very devastating move. You cannot move any of your other pieces until all your pieces are off the bar. The difficulty of entering the board from the bar varies greatly with how many safe points your opponent has established in his inner table. This should be a consideration when deciding to leave your own blot exposed to your opponent.

Example 2.8.1. *Suppose your opponent has four safe points in her inner table. What is the probability that you will be able to enter from the bar?*

Solution: With four safe points, only two points are available for your

TABLE 2.8:   Probabilities of Hitting a Single Blot with One Piece

No. of points away	No. of ways to hit	Prob. of hit
1	11	11/36
2	12	12/36
3		
4		
5		
6		
7		
8		
9		
10	3	3/36
11	2	2/36
12	3	3/36
13, 14, 17, 19, 21, 22, 23	0	0
15, 16, 18, 20, 24	1	1/36 for each

entry. Using Table 1.2, we see that this means there are only 20 rolls available to you, 11 for the first point and 11 for the second point, but two of these are the pairs that include both these points. Thus, your total is

$$2 \times 11 - 2 = 20.$$

Hence, the probability you escape the bar is $20/36 = 5/9$.   □

TABLE 2.9:   Probabilities of Entering from the Bar

No. of opponent's safe points	Probability of entering on next roll
0	1
1	
2	
3	3/4
4	5/9
5	
6	0

Example 2.8.2. *Suppose your opponent has only three safe points in*

her inner table. Now what is your probability of entering the board from the bar?

Solution: Now there are three points that you can use to enter the board. As each has 11 possible rolls containing that value, but again any two die values share two pairs, we see that there are a total of

$$33 - \binom{3}{2} \times 2 = 27$$

rolls that allow us to enter the board. Hence, our probability of entering is $27/36 = 3/4$. That is relatively high, even with half the points unavailable to us. This fact surprises many people. □

2.8.3 Bearing Off

You cannot bear off any of your pieces until all your pieces reach your inner table. You may bear off a piece with an exact roll (if such is possible) or with a larger roll if all your pieces are closer than the value rolled.

Bearing off is often a straightforward race situation. But there are instances when there is some critical strategy to bearing off and the good player should be aware of these opportunities. Common strategy is to bring pieces into your inner table with as economical a roll as possible. Such a strategy often means your 6 point becomes heavily loaded with pieces. Once all pieces are in the inner table, common strategy says to take as many pieces off as possible with each roll. Strategy really comes into play when there are just a few pieces left and your roll allows you to move closer, but not necessarily bear off. Some examples will help demonstrate what we mean.

Example 2.8.3. *Your inner table is as shown in Figure 2.5. You roll a 2 and a 1. What moves should you make?*

Solution: Since you have pieces on points 6 and 4 you have several options. You can end with pieces on points 5 and 2, or with pieces on points 4 and 3, or with pieces on points 6 and 1. Which we choose should be based on our next roll. By that we mean that we should position ourselves with the best chance to bear both pieces off on the next move. If we count the moves that will accomplish this goal, we can determine our probability of winning on the next roll.

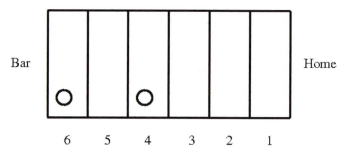

FIGURE 2.5:   Example 2.8.3 end game position.

From positions 6 and 1 there are a total of 11 rolls that contain a 6. But we can also bear off both pieces with double 2, 3, 4, or 5. Thus, there are a total of 15 favorable rolls. (Note, we can bear off from the 1 point with any roll of 1 or larger.) From the 5 and 2 points, we need rolls that contain a 5 or 6 but do not contain a 1. There are 20 rolls that contain a 5 or 6 and 4 of these contain a 1, namely $(1, 5), (5, 1)$ and $(1, 6)$ and $(6, 1)$. But double 4, 3, or 2 is favorable. Thus, we see that there are 19 favorable rolls from positions 5 and 2, so our probability of winning on the next roll would be 19/36. Finally, we consider the 4 and 3 points. Here we need a roll in which each number is at least 3 (as double 3 is fine). Thus, any roll containing a 1 or a 2 would be unfavorable, except double 2. There are 11 rolls with a 1 and 11 rolls with a 2, but 2 of these contain both. Hence, there are 19 unfavorable rolls, and thus only 17 favorable rolls, so our probability here is 17/36. From this we conclude that the best move is to positions 5 and 2.   □

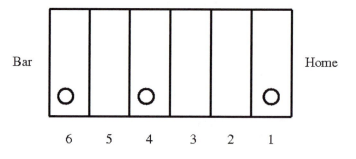

FIGURE 2.6:   Example 2.8.4 end game position.

Example 2.8.4. *You have three pieces left, on the points* 1, 4 *and* 6 *(see Figure 2.6). You now roll a double* 1. *Clearly you can only remove one piece from the board, but what positions do you leave for the next roll?*

Solution: Your choices are to leave your two remaining pieces at points 6 and 1, or 5 and 2, or 4 and 3. But these are the same as in the previous example. □

Example 2.8.5. *You again have three pieces left, on points* 3, 4 *and* 6 *(see Figure 2.7). You roll a double* 2. *What is your best move?*

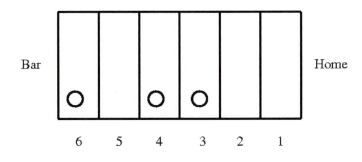

FIGURE 2.7: Example 2.8.5 end game position.

Solution: You have four moves of two points each. You can bear off the piece on the 6 point, and leave the other two at points 3 and 2; or you can bear off the piece on the 6 point and leave the other two at points 4 and 1; or you can bear off the piece on the 4 point, and leave the other two at 4 and 1 or 3 and 2 (so nothing new there).

But counting as we did before, from the 3 and 2 points we will finish with any roll where each die is at least 2. So we are only hurt by a roll containing at least one 1. There are thus 11 unfavorable rolls, so 25 favorable rolls. From the 4 and 1 points, we are only hurt by rolls where both numbers are 3 or less and not doubles, except double 1. Thus, the unfavorable rolls are $(2,1), (1,2), (3,1), (1,3), (3,2), (2,3)$ and $(1,1)$. Hence, there are 29 favorable rolls. Considering all the cases, we should prefer the 4 and 1 position. □

2.8.4   Doubling

The game of backgammon changed in 1925 with the introduction of the doubling cube (see [32]). After the game has started (with its initial bet) any player, upon beginning his turn, can double the stakes by flipping the cube to 2, thus indicating his intention of doubling the bet. The opposing player must then either concede the game (and the stakes prior to the double); or accept the doubling of the stakes and control of the cube (that is, they are the only ones who can next double). Clearly, such a move should be done only when there is a strong reason to do so, and we shall strive to discover that reason.

To understand the doubling cube, we will consider several cases.

1. When should I double?

2. When should I want my opponent to accept my double?

3. When should I accept a double from my opponent?

To answer these questions we need the ability to at least estimate the probability p of winning the game. Note that there are ways to do this. This is often done by counting the total number of points remaining to be traversed by us and by our opponent. Initially each player has 167 points to traverse. The interested reader should see [22].

We shall assume we have this ability. Given our (estimated) probability of winning, with $q = 1 - p$ the probability of losing, and an initial bet of s, we can consider these questions in terms of the expected value of doubling or not doubling and the options available to our opponent.

Example 2.8.6. *When should we consider doubling?*

Solution: Again we consider expected values.

$$EV(\text{no double}) = p(s) + (1 - p)(-s) = 2ps - s.$$

While, the expected value of doubling (when accepted) is

$$EV(\text{accepted double}) = p(2s) + (1 - p)(-2s) = 4ps - 2s.$$

Thus, we wish to know when

$$4ps - 2s \geq 2ps - s$$

which implies that

$$p \geq 1/2.$$

Hence, when $p = 1/2$ we conclude that either strategy is equivalent, while if $p > 1/2$, then doubling seems a good idea; while if $p < 1/2$ doubling would seem to be a bad idea. However, the player should also recognize that this does not incorporate the stage of the game. Doubling when p just exceeds $1/2$ early in a game may not be the best idea. While late in a game it is a far more powerful move. $\square$

Example 2.8.7. *Under what conditions do we want our double to be accepted?*

Solution: The expected value of doubling (when the double is accepted) remains $4ps - 2s$. While if the double is refused, the expected value becomes

$$EV(\text{refused double}) = s.$$

We then ask for

$$4ps - 2s \geq s$$

which implies that

$$p \geq 3/4.$$

Thus, we conclude that if our probability of winning is $3/4$ we do not care if the double is accepted or not. If $p < 3/4$ we would prefer a concession and if $p > 3/4$ we hope our opponent is willing to accept our double. $\square$

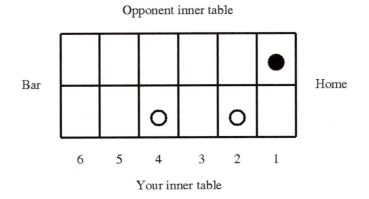

Opponent inner table

Bar   Home

6   5   4   3   2   1

Your inner table

FIGURE 2.8: An end game position, do you double?

Example 2.8.8. *You are white and the board is as shown in Figure 2.8. It is your roll and you control the doubling cube. Should you double?*

Solution: To answer this question we must determine our probability of winning on the next roll, since if given a chance, our opponent will certainly win. We have two pieces remaining in positions 2 and 4. Thus, we will win unless our roll contains a 1 and we know there are 11 ways that could happen, or we roll a $(2, 3)$, or $(3, 2)$. Thus, there are a total of 13 unfavorable rolls. Hence, our probability of winning on the next roll is $23/36 = .6389 > .5$ and we should double.   □

Example 2.8.9. *You are again white and the board is now as shown in Figure 2.9. It is your roll and you control the doubling cube. Should you double?*

Opponent inner table

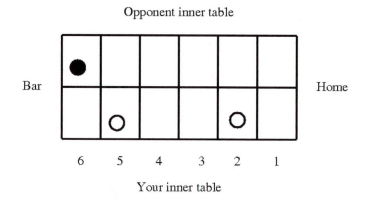

Your inner table

FIGURE 2.9: Another end game position, do you double?

Solution: You might immediately say that we should double as this position seems even better than in the previous example. But there are more things to consider here. We might not win on our first roll and our opponent also might not win on her first roll. But we almost certainly would face a double from our opponent, if he or she have the chance. Also, our own position is slightly worse than in the previous example. The events of interest are thus:

- $E_1 =$ you win on your first roll,

- E_2 = your opponent wins on her first roll,

- E_3 = you win on your second roll (not completely guaranteed),

- E_4 = your opponent wins on her second roll (this is guaranteed).

So what are the relevant probabilities? We see that $P(E_1) = 19/36$ as we need rolls which contain a 5 or 6 and also do not contain a 1, or a roll of double 2, 3 or 4. Event E_2 can only happen if E_1 fails. In addition, there are only 9 unfavorable rolls for your opponent. Thus, $P(E_2) = 17/36 \times 3/4$. Next we see that event E_3 can only happen if both E_1 and E_2 fail to occur. This happens with probability $17/36 \times 1/4$. Thus $P(E_3) = 17/36 \times 1/4 \times 34/36$, as there are only two rolls that are unfavorable. Finally, $P(E_4) = 1$, provided we reach that stage. This happens with probability $17/36 \times 1/4 \times 2/36$. Now we can compute expected values. We assume a bet of 1 unit for simplicity.

$$EV(\text{no double}) = \frac{19}{36}(1) + \frac{17}{36} \times \frac{3}{4}(-1)$$

$$+\frac{17}{36} \times \frac{1}{4} \times \frac{34}{36}(1) + \frac{17}{36} \times \frac{1}{4} \times \frac{2}{36}(-1)$$

$$= 0.2785$$

$$EV(\text{double, accept redouble}) = \frac{19}{36}(2) + \frac{17}{36} \times \frac{3}{4}(-4)$$

$$+\frac{17}{36} \times \frac{1}{4} \times \frac{34}{36}(4) + \frac{17}{36} \times \frac{1}{4} \times \frac{2}{36}(-4)$$

$$= 0.0586$$

$$EV(\text{double, refuse redouble}) = \frac{19}{36}(2) + \frac{17}{36}(-2) = \frac{4}{36} = 0.1111.$$

Comparing these expectations we can see that the best course is to hold off on doubling because of the threat of a redouble that greatly changes the situation.

Even if we now ignore the slim chance that we roll $(2, 1)$ two times in a row, we find that the three remaining events have probabilities

$$P(\text{win 1st roll}) = 19/36$$

$$P(\text{opponent wins}) = 17/36 \times 3/4$$

$$P(\text{win on 2nd roll}) = 17/36 \times 1/4.$$

Now the expected value computations yield

$$EV(\text{no double}) = 21/72$$

$$EV(\text{double, opp. accepts}) = 8/72$$

$$EV(\text{double, refuse redouble}) = 8/72.$$

Thus, again we see the decision should be to hold off on doubling your opponent.   □

Exercises:

2.8.1 Complete Table 2.8.

2.8.2 Complete Table 2.9.

2.8.3 Determine the conditions under which you should accept a double from your opponent.

2.8.4 In backgammon a player's count on a given roll is the total number of points that his pieces may move. Compute the expected count in backgammon.

2.8.5 Your opponent has left two blots 3 points and 6 points from one of your pieces. What is the probability you can hit your opponent on your next roll?

2.8.6 Suppose conditions are as in the previous problem, but your opponent has also established a safe point 4 points from your piece. Now what is the probability you can hit one of his blots on your next roll?

2.8.7 Conditions are as in problem 6, but the safe point is 2 points away from your piece. Now what is the probability you can hit your opponent on your next roll?

2.8.8 You have three pieces on the first point of your inner table and one piece on the second point. Your opponent has two pieces on his first point and one on his third point. It is your roll. What should you do?

2.8.9 You are on the bar. Your opponent has established safe points on points $1, 2$ and 3. But your opponent has blots on points 4 and 6. What is the probability of hitting a blot on your roll? What is the probability of entering the table, but not hitting a blot?

2.8.10 Suppose you have two pieces on the bar and your opponent has established 2 safe points in his inner table. What is the probability you can enter both pieces on your next roll?

2.8.11 Suppose the doubling cube is replaced by a tripling cube (that is, the stakes are tripled). What is the probability above which you should triple your opponent?

2.8.12 Your opponent has only two pieces remaining on the first point of her inner table. You have only two pieces on your 3 and 4 points of your inner table. What should you do?

2.8.13 Suppose the situation is as in the previous problem except your pieces are on the 2 and 5 points. Now what should you do?

Chapter 3

Repeated Play

3.1 Introduction

As stated earlier, repetition is the key idea in using probability as a tool. Expectation provides a way of predicting the long-term average of repeating an experiment (or playing a game). But is there even more we can learn?

In this chapter we will concentrate on repeated play in fixed probability games. This is an attempt to learn more about the possible outcomes. By repeated play we do not mean playing a game two, three or even eight times, but rather playing dozens to hundreds or even thousands of times. What we seek is information about how the outcomes are distributed, how likely are (winning or losing) streaks to occur, what long-term strategies people tend to adopt, why these strategies tend to fail, and finally, what are our real chances of breaking the house? We can use all these bits of information to answer more and different kinds of questions. Not only just how likely we are to "beat the odds" if we make a large number of bets of a certain kind, but much more about what to expect should we attempt this.

Repeated play allows us a statistical look at these games. By repeatedly playing a game, the outcomes become a statistical *sample population*, that is, a subset of all possible outcomes. (Note that the set of all outcomes is often impossible to generate). We shall take some advantage of this in answering our questions. Note that for the sake of examples, we shall assume that repeated sports actions like at bats in baseball or shots in basketball fall into the category of independent fixed probability events.

3.2 Binomial Coefficients

Earlier we defined the *binomial coefficient* $\binom{n}{r}$ as the number of ways of choosing r elements from an n element set, when order does not matter. Now we want to see that this computation arises even more often than we had seen earlier (and it arose often then)!

To this end, suppose we ask the following question:

Question 3.2.1. *What is the probability of getting exactly three tails if I flip a fair coin eight times?*

Solution: To answer this question we should think about the sequence of experiments (coin flips). One such sequence is

$$H\ T\ H\ H\ T\ H\ T\ H$$

and there are exactly three tails in this sequence. But there are many other such sequences. Our job is really to count all such sequences, as this will be fundamental in answering our question.

When considering the sequence above, we can use the multiplication principle to compute the probability of this one sequence happening. It is just the product of the probabilities of the individual events in the sequence. Thus, the probability is

$$\frac{1}{2} \times \frac{1}{2} \times \frac{1}{2} \times \frac{1}{2} \times \frac{1}{2} \times \frac{1}{2} \times \frac{1}{2} \times \frac{1}{2} = (\tfrac{1}{2})^5(\tfrac{1}{2})^3 = (\tfrac{1}{2})^8 = \frac{1}{256}$$

as each individual outcome has probability $1/2$. The term $(\tfrac{1}{2})^5$ represents the probability of obtaining the five H (heads) in the sequence and the term $(\tfrac{1}{2})^3$ represents the probability of obtaining the three T (tails).

Note that the final computation would be the same for any sequence of eight flips containing exactly three tails or not, since the probability of a tail or head is .5. To obtain the overall desired probability of the event of having exactly three tails in eight flips of a fair coin, we need to multiply the above probability (which holds for each sequence of length eight) by the number of disjoint, equally likely ways of obtaining a sequence of length eight with exactly three tails. In order to accomplish our goal we need to place exactly three tails in a sequence of eight coin flips. This is equivalent to selecting the three positions in the sequence for the T (tails). But we know there are exactly $\binom{8}{3}$ ways to do this. Thus, our computation would be:

$$P(\text{exactly 3 tails}) = \binom{8}{3} \times (\frac{1}{2})^5 \times (\frac{1}{2})^3 = \frac{7}{32}.$$

$\square$

This now raises an even larger question.

Question 3.2.2. *Can we determine the probability of obtaining exactly r tails in eight flips of a fair coin, for each r in the range $0 \leq r \leq 8$?*

Solution: In order to answer this question, we wish to take advantage of what we have just learned! Our solution will be the number of ways of placing exactly r tails in a sequence of eight flips multiplied by the probability of obtaining any one of these sequences. Again, each of these sequences is equally likely because the probability of a head or tail is .5. In general there are r tails and $8 - r$ heads in the sequence. Hence, we can write a general solution as:

$$P(\text{exactly } r \text{ tails}) = \binom{8}{r} \times (\frac{1}{2})^r \times (\frac{1}{2})^{8-r}. \quad \square \qquad (3.1)$$

This general formula for our experiment allows us to build Table 3.1. We should stop and notice several things about the experiment we have just performed. Equation 3.1 can easily be generalized to any number of flips of a coin. If we were to flip a coin n times, then

$$\binom{n}{r} \times (\frac{1}{2})^r \times (\frac{1}{2})^{n-r} \qquad (3.2)$$

would give the probability of exactly r tails, $0 \leq r \leq n$.

But we can take this even further. The important point here is that flipping a coin is a fixed probability experiment. Given any fixed probability experiment with probability p of success (whatever success might be) and probability $q = 1 - p$ of failure, if we perform this experiment n times, then the probability of exactly r successes in n tries is:

$$\binom{n}{r} \times p^r \times q^{n-r}. \qquad (3.3)$$

The argument why this is the probability follows exactly as our earlier argument on coin flips. There are $\binom{n}{r}$ ways to create a sequence of experiments with exactly r successes in n tries. Within any sequence,

TABLE 3.1:   Probabilities for Exactly r
Tails in Eight Flips of a Fair Coin

No. Tails	No. Ways	Probability
0	$\binom{8}{0} = 1$	$\frac{1}{256}$
1	$\binom{8}{1} = 8$	$\frac{8}{256}$
2	$\binom{8}{2} = 28$	$\frac{28}{256}$
3	$\binom{8}{3} = 56$	$\frac{56}{256}$
4	$\binom{8}{4} = 70$	$\frac{70}{256}$
5	$\binom{8}{5} = 56$	$\frac{56}{256}$
6	$\binom{8}{6} = 28$	$\frac{28}{256}$
7	$\binom{8}{7} = 8$	$\frac{8}{256}$
8	$\binom{8}{8} = 1$	$\frac{1}{256}$

each of these r successes happens with probability p and each of the failures happens with probability q. Thus, any one of the sequences happens with probability $p^r \times q^{n-r}$. We then multiply by the number of such sequences which is $\binom{n}{r}$, producing Equation 3.3.

Now we may ask other questions of a similar nature.

Example 3.2.1. *In the World Series, suppose the probability the Red Sox will defeat the Cardinals in any one game is .6. What is the probability that the World Series will go seven games?*

Solution: We first note that our assumptions here make this a fixed probability experiment (even though in real life it would not be so simple, we will accept this for now). The first team to win four games wins the Series. Thus, for the World Series to go seven games requires that the Red Sox win exactly three of the first six games (and thus the Cardinals also win three of the first six games). This question fits exactly our general pattern. Hence, the probability that the Red Sox win exactly three of the first six games is:

$$\binom{6}{3} \times (.6)^3 \times (.4)^3 = .2765.$$

□

Now we ask a slightly harder question.

Question 3.2.3. *What is the probability the Red Sox will win the series in five games?*

Solution: To answer this question we must be careful. We must have the Red Sox win the 5*th* game and have this be their 4*th* victory, in order to have the World Series end after five games.

Thus, the Red Sox must win exactly three of the first four games and also win game five. These are two independent events and so we can handle this situation. The probability the Red Sox win three of the first four games is:

$$\binom{4}{3} \times (.6)^3 \times (.4) = .3456$$

and thus, the probability the Red Sox win the Series in five games is:

$$.3456 \times .6 = .2074.$$

□

More examples will help us see other facts about these types of questions.

Question 3.2.4. *The National Heads Up Poker Tournament pits players against one another in an elimination tournament. The final two players play a best two out of three matches for the championship. Suppose Sam Farha has a probability of .55 of winning any one of these matches against his less-experienced opponent.*

  1. *What is the probability all three matches are required?*

  2. *What is the probability Farha wins in two matches?*

Solution: This follows in a manner similar to what we have already seen.

(1) For three matches to be played, the players must split the first two matches. Thus,

$$P(3 \text{ matches}) = \binom{2}{1} \times (.55) \times (.45) = .495.$$

(2) For Farha to win in two straight matches we have

$$P(2 \text{ matches}) = \binom{2}{2} \times (.55)^2 \times (.45)^0 = .3025.$$

$\square$

Binomial coefficients are closely linked to the coefficients of powers of $(p + q)$ (hence the name — the coefficients in an expansion of a power of a binomial expression). From basic algebra we know that

$$(p + q)^2 = 1p^2 + 2pq + 1q^2$$

and

$$(p + q)^3 = 1p^3 + 3p^2q + 3pq^2 + 1q^3.$$

Here we note that $\binom{n}{0} = \binom{n}{n} = 1$ while $\binom{2}{1} = 2$ and $\binom{3}{1} = \binom{3}{2} = 3$. Thus, we could write $(p + q)^2$ as

$$(p + q)^2 = \binom{2}{0}p^2 + \binom{2}{1}pq + \binom{2}{2}q^2$$

and $(p + q)^3$ as

$$(p + q)^3 = \binom{3}{0}p^3 + \binom{3}{1}p^2q + \binom{3}{2}pq^2 + \binom{3}{3}q^3.$$

In fact, in general we have the following well-known and important theorem.

Theorem 3.2.1. The Binomial Theorem. *For any numbers p and q and any positive integer n*

$$(p + q)^n = \binom{n}{0}p^n + \binom{n}{1}p^{n-1}q + \binom{n}{2}p^{n-2}q^2 + \ldots + \binom{n}{n}q^n. \quad (3.4)$$

There are several ways to prove the Binomial Theorem. If you are familiar with the technique of induction, it offers a very straightforward proof. Instead, we shall argue directly.

Proof of Binomial Theorem: Consider

$$(p+q)^n = (p+q)(p+q)...(p+q).$$

In multiplying this product, one obtains 2^n terms by selecting either a p or a q from each of the n factors. Each of these terms is made up of n such factors, some of them being p's and the rest q's. Say there are k p's and $n - k$ q's. Certain of these terms will be alike and can be grouped together. Namely, there will be as many terms with k p's as there are ways of selecting k out of the n terms. Thus, the coefficient of $p^k q^{n-k}$ will be $\binom{n}{k}$, and the result follows. □

Example 3.2.2. *Suppose you play a game where the probability of winning is $\frac{2}{5}$. Find the probabilities for the random variable $X = k$, where X represents the number of wins $(k = 0, 1, \ldots, 5)$ in five plays of the game.*

Solution: To find the probabilities for X we use $p = \frac{2}{5}$ and $q = \frac{3}{5}$. Then the terms for the various probabilities are represented by the terms of $(p + q)^5$. By the Binomial Theorem we see that

$$(2/5 + 3/5)^5 = (2/5)^5 + 5(2/5)^4(3/5) + 10(2/5)^3(3/5)^2$$
$$+ 10(2/5)^2(3/5)^3 + 5(2/5)(3/5)^4 + (3/5)^5.$$

Thus, for example we see that the probability of exactly two wins is represented by the term $10(2/5)^2(3/5)^3 = .3456$ while the probability of three or more wins is represented by the sum

$$(2/5)^5 + 5(2/5)^4(3/5) + 10(2/5)^3(3/5)^2 = .1761.$$

The probability of exactly four wins is represented by

$$5(2/5)^4(3/5) = .0768$$

and so forth. □

Next we consider a somewhat different look at the Binomial Theorem. This is the well-known Pascal's Triangle.

Using the Binomial Theorem (Theorem 3.2.1), we can see that

$$(p + q)^0 = 1$$
$$(p + q)^1 = 1p + 1q$$

$$(p+q)^2 = 1p^2 + 2pq + 1q^2$$

$$(p+q)^3 = 1p^3 + 3p^2q + 3pq^2 + 1q^3$$

$$(p+q)^4 = 1p^4 + 4p^3q + 6p^2q^2 + 4pq^3 + 1q^4$$

$$(p+q)^5 = 1p^5 + 5p^4q + 10p^3q^2 + 10p^2q^3 + 5pq^4 + 1q^5$$

and so forth.

Pascal understood all this and saw the relationships between these coefficients. He used this to create a way to compute them more easily (at least when they are small).

TABLE 3.2:   Pascal's Triangle

$$
\begin{array}{ccccccccccccccc}
 & & & & & & & 1 & & & & & & & \\
 & & & & & & 1 & & 1 & & & & & & \\
 & & & & & 1 & & 2 & & 1 & & & & & \\
 & & & & 1 & & 3 & & 3 & & 1 & & & & \\
 & & & 1 & & 4 & & 6 & & 4 & & 1 & & & \\
 & & 1 & & 5 & & 10 & & 10 & & 5 & & 1 & & \\
 & 1 & & 6 & & 15 & & 20 & & 15 & & 6 & & 1 & \\
1 & & 7 & & 21 & & 35 & & 35 & & 21 & & 7 & & 1 \\
\end{array}
$$

$$\vdots$$

In considering Table 3.2, note that each line begins and ends with a 1, as the leading and final coefficients in the binomial theorem are $\binom{n}{0}$ and $\binom{n}{n}$, each of which has value 1. The other terms on each line, after the second line, are created as the sum of the terms in the line above that lie directly to the left and right of the term we are creating. This pattern continues. Thus, we may determine the coefficients for an expansion of $(p+q)^r$ by just finding the appropriate line of the triangle. Line one represents the coefficients when $r = 0$, line two corresponds to $r = 1$, and so forth.

Example 3.2.3. *Let two coins be tossed simultaneously six times. What is the probability that double heads will appear four or more times?*

Solution: We seek the probability that double heads appears four, five, or six times. We know that the probability of double heads happening on any one toss is $\frac{1}{4}$. Thus, by the Binomial Theorem, the probability we seek is given by the last three terms in the polynomial expansion of

$$\left(\frac{3}{4} + \frac{1}{4}\right)^6.$$

Hence, using Pascal's triangle we see this sum is just

$$15\left(\frac{3}{4}\right)^2 \left(\frac{1}{4}\right)^4 + 6\left(\frac{3}{4}\right)^1 \left(\frac{1}{4}\right)^5 + \left(\frac{1}{4}\right)^6 = \frac{154}{4^6} = .0376.$$

Note that these values are just the same terms we would obtain if we had selected exactly four, five and six double heads to occur, computed them separately and then added these terms together. That is,

$$\binom{6}{4}\left(\frac{3}{4}\right)^2\left(\frac{1}{4}\right)^4 + \binom{6}{5}\left(\frac{3}{4}\right)\left(\frac{1}{4}\right)^5 + \binom{6}{6}\left(\frac{1}{4}\right)^6.$$

$\square$

Now suppose we ask for the probability you will win 250 times in 500 bets on red at roulette. Then, by our rules this probability is

$$\binom{500}{250} \times \left(\frac{18}{38}\right)^{250} \times \left(\frac{20}{38}\right)^{250}.$$

Although it was easy to write down the expression for this probability, the fact that $n = 500$ makes the exponents large and thus makes actually computing this value by hand difficult.

Question 3.2.5. *Can we find a way around this problem?*

Exercises:

3.2.1 Five cards are drawn from a standard deck. After each card is drawn, it is replaced in the deck and the deck is shuffled. This ensures the probabilities are the same for each drawing. Create a table showing the probability 0, 1, 2, 3, 4 or 5 spades are drawn from the deck.

3.2.2 Using Exercise 3.2.1, determine the expected number of spades drawn from the deck in five drawings. What is the variance for this drawing game?

3.2.3 A die is rolled three times and each time we record whether the number shown is odd (O) or even (E). Let the random variable X be the total number of evens recorded. Find the probability distribution for $P(X)$.

3.2.4 Assume when a deer hunter shoots at a deer, the probability of hitting the deer is .6. Find the probability that the hunter:

1. Will hit exactly 4 of the next 5 deer at which he shoots.

2. Will hit at least 3 of the next 5.

3. Will hit at least 1 of the next 5.

3.2.5 It is known that 60% of the fans in Philadelphia are opposed to a particular player being traded. If 20 fans are selected at random, find the probability (as an expression only) that:

1. Exactly 10 are opposed to the trade.

2. More than 17 are opposed to the trade.

3. Fewer than 3 are opposed to the trade.

3.2.6 Suppose 18% of the players in the NBA are foreign born. If six players are selected at random, what is the probability that:

1. Exactly one is foreign born.

2. Two or three are foreign born.

3. More than four are foreign born.

3.2.7 The probability a bridge hand has no aces is .3. What is the probability that a player will be dealt no aces three times in four hands?

3.2.8 Hospital records indicate that 40% of the baseball pitchers that have a certain tendon surgery never pitch again. If six pitchers all have this surgery, find the probability at least four will pitch again.

3.2.9 A baseball player has a batting average of .300. Assuming this represents his probability of getting a hit in any at bat, find the probability that in his next five at bats he gets:

1. At most one hit.

2. Exactly two hits.

3. Exactly three hits.

4. At least three hits.

3.2.10 If a basketball player makes free throws at an 80% rate, what is the probability he will make:

1. Seven free throws in his next eight tries?

2. Five free throws or less in his next eight tries?

3.2.11 If a hockey player has a probability of .5 of receiving a penalty in any game, what is:

1. The probability the player goes four games without a penalty?

2. The probability the player receives a penalty in three of the next four games?

3.2.12 Two evenly matched tennis players play four sets of tennis. Is the score more likely to be $3 - 1$ or $2 - 2$? Suppose they played six sets. Is the score more likely to be $3 - 3$ or $4 - 2$?

3.3 The Binomial Distribution

Our goal in this section is to answer Question 3.2.5. In doing this we will settle for very good approximations of the probabilities we seek, especially if it allows us to avoid messy computations (where we would only get approximations anyways).

Consider the probabilities shown in Table 3.1. Let X be the random variable representing the number of tails occurring in the eight flips of the coin. Then the probability $P(X = i)$ is shown for each $i = 1, 2, \ldots, 8$. We call these values a *probability distribution for X*. We can graph this distribution as a bar graph, (sometimes called a *histogram*) with each rectangular bar centered on i with width 1 and height $P(X = i)$. Thus, each bar has area $P(X = i)$. We show this graph in Figure 3.1, with the areas of each rectangle shown below the rectangle.

If we consider the probability distribution for the number of tails when we flip a coin two times and four times, we see that the general shape of the graphs are very similar. (See Figures 3.2 and 3.3.)

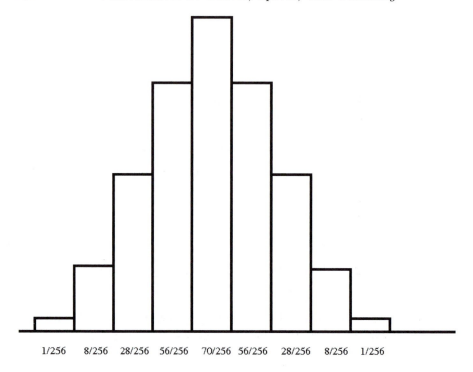

FIGURE 3.1:   Probabilities of i tails in eight flips.

If we continued this pattern with 16 flips, 32 flips, 64 flips, etc., we would see that the general shape of the graphs continues to be a bell shape and the larger the number of flips, the closer to a bell shape the graph would appear. Thus, the more we repeat the experiment of flipping the coin, the better a bell curve approximates our experiment.

Example 3.3.1. *What is the probability that the number of tails is at most five if we flip a fair coin eight times?*

Solution: From Table 3.1 we can see that this is just the sum of the first six values in the table. That is,

$$\frac{1}{256} + \frac{8}{256} + \frac{28}{256} + \frac{56}{256} + \frac{70}{256} + \frac{56}{256} = \frac{219}{256} = .8557.$$

This was a bit tedious, and if the question had been what is the probability of 50 or fewer tails in 80 flips of a fair coin we would not want to do this by hand. Happily, there is a better way.

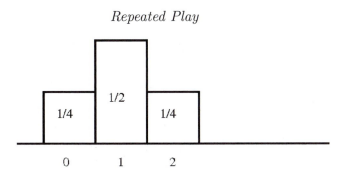

FIGURE 3.2:   Probability distribution for two flips.

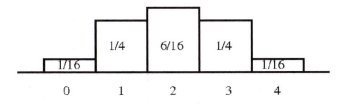

FIGURE 3.3:   Probability distribution for four flips.

The coin flip is only one of many *binomial experiments* (i.e., two outcomes) we might consider. These various experiments have different probabilities of success and hence somewhat different probability distributions. We seek a way to approximate any of these.

Figure 3.4 shows an approximating bell curve for the probability distribution of Table 3.1. As the number of repetitions of an experiment increase, such curves become better and better approximations of the distribution. These bell-shaped normal distribution approximating curves are obtained by graphing exponential functions of the form $f(x) = ae^{-c(x-\mu)^2}$ where the constants a, μ, c are chosen so that:

1. The total area under the curve $f(x)$ is 1 (the total probability).

2. The peak of the curve is at the mean ($\mu = np$) on the horizontal axis.

3. The overall shape of the bell reflects the standard deviation of the distribution.

The one approximating curve we seek is called the *standard normal distribution* . The standard normal distribution is one where the mean

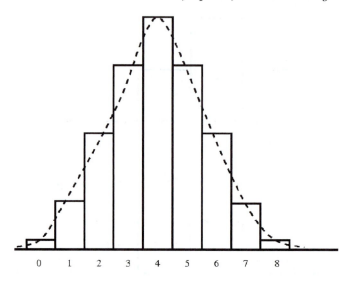

FIGURE 3.4: Approximating bell curve for probability distribution.

$\mu = 0$ and standard deviation is 1.

Using this one curve, a table of probabilities can be compiled so that we can look up answers to various questions and obtain very good approximations to the massive computations we wish to avoid. Table 9.1 is one such table. It provides the probability $P(r \leq Z)$ for values of Z in the range 0.0 to 3.09. This is enough of a range to answer all our questions. For values of $Z \geq 3.1$ we may assume that $P(Z) = 1$ and for negative values of Z we will find a corresponding positive Z (usually just $|Z|$) to use instead. Examples of what to do in such situations will be covered in the examples that follow.

One test needs to be applied to be sure we will get a good approximation using the standard normal distribution. We only use Table 9.1 provided

$$np > 5 \text{ and } nq > 5.$$

Now, if we seek the probability of 50 or fewer tails in 80 flips of a fair coin, we could follow our earlier examples and simply add together the terms corresponding to $r = 0, 1, \ldots, 50$ from the expansion of $(1/2 + 1/2)^{80}$, or alternately, we could seek the aid of Table 9.1. There is a straightforward conversion formula for finding the appropriate value to look up in the table. If we seek the probability for at most r successes in n attempts of an experiment with some binomial probability distribution, then

$$Z_r = \frac{(x - .5) - \mu}{\sigma} = \frac{(x - .5) - np}{\sqrt{npq}}$$

is the appropriate value to look up in the standard normal distribution table (Table 9.1).

Thinking of this formula as

$$Z = \frac{\hat{x} - \mu}{\sigma}$$

for some value $\hat{x}$ of interest, allows us to understand what this mapping is accomplishing. Each value of $\hat{x}$ determines a value of Z, so Z itself is a random variable. Notice that this definition implies that if $\mu = 90$, $\sigma = 10$ and we plug in $\hat{x} = 100$, then

$$Z = \frac{100 - 90}{10} = \frac{10}{10} = 1.$$

This corresponds to the fact that $\hat{x}$ is one standard deviation above the mean. If instead $\hat{x} = 70$, then

$$Z = \frac{70 - 90}{10} = \frac{-20}{10} = -2.$$

This corresponds to the fact that 70 is 2 standard deviations below the mean. Since the standard normal curve has mean equal to zero, Z corresponds to the point where the area under the standard normal curve equals that of the distribution of interest.

Some examples should help make clear these ideas.

Example 3.3.2. *Find the probability of at most 50 tails in 80 flips of a fair coin.*

Solution: Since $(80)(.5) > 5$ we may use the table. Using the formula for $Z = Z_r$ we find that

$$Z_{50} = \frac{(50 - .5) - (80)(.5)}{\sqrt{(80)(.5)(.5)}} = \frac{49.5 - 40}{\sqrt{20}} = \frac{9.5}{4.472} = 2.12.$$

Now from Table 9.1 the probability corresponding to $r \leq 50$ and thus $Z_{50} = 2.12$ is .9830. The corresponding area under the standard normal curve is shown in Figure 3.5. □

We now further illustrate these ideas with a series of examples.

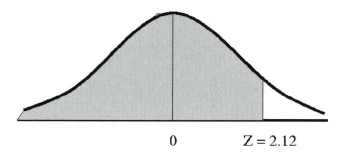

FIGURE 3.5:   Area under bell curve when $Z = 2.12$.

Example 3.3.3. *What is the probability you are ahead (winning money) after 100 bets of \$1 on red playing American Roulette?*

Solution: We know that American Roulette pays even money on a red bet and that the probability of red winning is $18/38 = .474$. Note that $(100)(.474) > 5$ and $(100)(.526) > 5$ and so we may use Table 9.1.

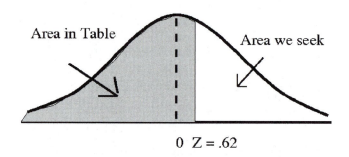

FIGURE 3.6:   We seek area to the right of $x = Z$.

Now, we need to compute the minimum number of wins that would allow us to be money ahead. Let r be this number of wins and hence $100 - r$ is the number of losses. Then, since we win one dollar r times and lose one dollar $(100 - r)$ times, the value we seek is the solution to

$$r + (-1)(100 - r) > 0.$$

or

$$2r > 100$$

Hence, we are money ahead provided $r \geq 51$. This makes perfect sense since to be money ahead in an even money betting situation, you must win more times than lose. Now the Z value that corresponds to $r = 51$ is

$$Z_{51} = \frac{(51 - .5) - (100)(.474)}{\sqrt{(100)(.474)(.526)}} = \frac{3.1}{\sqrt{24.93}} = .62.$$

We can look up $Z_{51} = .62$ in Table 9.1, but we seek the area to the right of this line (see Figure 3.6) and the table gives the area to the left of the line $Z = .62$. Thus, to obtain the proper value we subtract the table value from 1. Hence, the probability that corresponds to $r \geq 51$ is

$$1 - .7324 = .2672. \quad \square$$

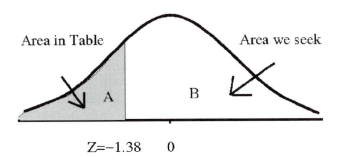

FIGURE 3.7:   Example of a Z score that is negative.

Example 3.3.4. *What is the probability you are less than $20 behind after* 100 *bets on red at American Roulette?*

Solution: We already know we can use the table in this situation. We must determine the number of wins that corresponds to this case. In this problem, this value is the solution to

$$r + (-1)(100 - r) > -20.$$

Hence, $r \geq 41$. The corresponding Z_{41} is

$$Z_{41} = \frac{(41 - .5) - (100)(.474)}{\sqrt{(100)(.474)(.526)}} = \frac{-6.9}{4.99} = -1.382.$$

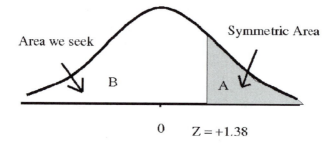

$$0 \qquad Z = +1.38$$

FIGURE 3.8:   Symmetry allows us to obtain wanted value.

But this Z is negative and our table only shows positive Z. Further, we also want the value to the right of $Z_{41} = -1.38$. Using the symmetry of the standard normal curve, this will correspond to the area to the left of $Z = +1.38$ (see Figures 3.7 and 3.8). The corresponding probability from Table 9.1 is .9162.   □

Example 3.3.5. *What is the probability you are $40 or more ahead after* 500 *$1 bets on a single number at American Roulette?*

Solution: Since $(500)(1/38) > 5$ and $(500)(37/38) > 5$ we may use the table. Using the House odds for a single number bet, the number of wins we seek is the solution to

$$35r + (-1)(500 - r) \geq 40.$$

Hence,

$$36r \geq 540 \text{ or } r \geq 15.$$

Thus, Z_{15} for $r \geq 15$ is

$$Z_{15} = \frac{(15 - .5) - (500)(1/38)}{\sqrt{(500)(1/38)(37/38)}} = \frac{1.342}{3.58} = .04.$$

Hence, the correct probability is $1 - .5160 = .4840$, as we want $r \geq 15$ to hold.   □

Example 3.3.6. *What is the probability you are between $50 behind to $50 ahead after* 200 *bets of $1 on a single number playing American Roulette?*

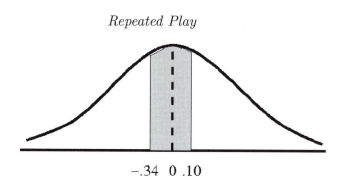

−.34 0 .10

FIGURE 3.9: Area for Example 3.3.6.

Solution: This time we must find two values to bound the range of "winnings." First we seek the minimum solution to

$$35r + (-1)(200 - r) \geq -50$$

and hence $r \geq 5$ as r must be an integer. We also seek the solution to

$$35r + (-1)(200 - r) \leq 50$$

and hence $36r \leq 250$ and so $r \leq 6$ (again as r must be an integer). So the solution to the question implies we must win 5 or 6 times in 200 tries to be in the desired range of "winnings." The corresponding Z values are

$$Z_5 = \frac{(5 - .5) - (200)(1/38)}{\sqrt{(200)(1/38)(37/38)}} = -.34.$$

While,

$$Z_6 = \frac{(6 - .5) - (200)(1/38)}{2.562} = .10.$$

The corresponding area under the standard normal curve is shown in Figure 3.9. Looking at this figure helps us understand the computations we must now perform.

To find the desired probability we find the probability for Z_6 and then subtract the probability for Z_5. Of course, to find the probability for $Z_5 = -.34$ we will instead use $+.34$ (recall the table gives the area to the left of the line). Thus, using Table 9.1 we see that this yields

$$0.5398 - (1 - 0.6331) = .1729. \quad \square$$

At this point we need to recognize some subtle points. In the previous section we computed precisely probability distributions for various binomial distributions. But in practice we perform experiments repeatedly (sometimes called obtaining *samples* from a *population* and obtain success or failure many times). We should not expect to obtain perfect probability distributions when we do this. For example, if we flip a coin 1000 times, having 495 heads and 505 tails is a reasonable outcome. It does not represent a perfect probability distribution, but it certainly is close. Expecting 500 heads and 500 tails is more unreasonable in practice.

But, there is a deep and fundamental result that tells us approximately what to expect when an experiment is repeated many times, that is, a large sample is obtained. It is called the *Central Limit Theorem*.

Theorem 3.3.1. Central Limit Theorem. *If a random sample of n observations is obtained from a population, then when n is sufficiently large, the sample distribution will be approximately a normal distribution. The larger the sample size, the better will be the normal approximation.*

This theorem is fundamental in statistics. It justifies the use of the standard normal curve method on a wide variety of problems. We cannot say exactly how large n must be for the theorem to apply, but usually $n \geq 30$ is sufficient. The closer the distribution being sampled is to a normal curve, the better will be the approximation, regardless of the size of n. Since we are primarily concerned with binomial distributions here, our earlier tests are sufficient for our purposes. Note that the statement of the Central Limit Theorem here is a somewhat simplified one.

Exercises:

3.3.1 You play a game with an even money payoff and a probability of winning each game played of $p = .48$. If the game is played 200 times, what is the probability you will be $10 or more behind?

3.3.2 Suppose you are playing blackjack, with a probability of winning any hand of .49. What is the probability you leave being $0 or more ahead after 100 hands?

3.3.3 What is the probability of being $100 or more behind after 1000 bets of $1 on a single number in American roulette?

3.3.4 What is the probability you are between $25 behind to $25 ahead after 400 bets on black at American roulette?

3.3.5 What is the probability you are more than $100 ahead after 300 bets of $5 on a single number in American roulette?

3.3.6 What is the probability you are at least $10 behind after 500 bets of $1 in a game where the probability of a win in any game is .4?

3.3.7 What is the probability you are $50 to $100 behind after 300 bets of $2 on red in American roulette? What is the probability you are money ahead after 100 games?

3.3.8 You play a game that pays 20:1 and your probability of winning is .06. What is the probability you are at least $5 ahead after 100 bets of $1?

3.3.9 Playing the game from the previous problem, what is the probability you are between $50 behind to $50 ahead after 200 bets of $1?

3.3.10 What is the probability you are no more than $30 behind after 300 bets on a single number in American Roulette?

3.3.11 What is the probability you are more than $10 behind playing the same game as in the previous problem?

3.3.12 You play a game with a 3:2 payoff and a probability of winning of .44. What is the probability you are ahead after 100 bets of $2? Explain why your answer makes sense.

3.4 The Poisson Distribution

In this section we wish to continue the exploration of repeated play. Along the way we will develop another distribution which approximates the binomial distribution in many cases. This distribution has many applications especially to problems related to describing the number of events that will occur in a specified period of time or in a specified area or range.

Our motivating example is again roulette. Recall that in American roulette there are 38 numbers. From what we know about independent events, it seems hard to expect that in 38 spins of the roulette wheel, all 38 distinct numbers will appear exactly once.

To convince ourselves of this, note that there are 38^{38} possible 38 number sequences (the events we are considering). All 38 numbers appearing in 38 spins means we have a permutation of the 38 numbers. We know there are 38! permutations. Thus, the probability that all 38 numbers appear in 38 spins of the wheel is

$$P(\text{all 38 appear }) = \frac{38!}{38^{38}} = 4.861203 \times 10^{-16}$$

$$= .0000000000000004861203.$$

You are more likely to be hit by lightning and then win the lottery than you are to see all 38 numbers appear in 38 spins of the roulette wheel. Since we cannot expect to see all 38 distinct numbers, our fundamental question becomes:

Question 3.4.1. *How many different numbers should we expect to appear on average in* 38 *spins of the roulette wheel?*

Our question really asks:

Question 3.4.2. *What is the expected value of the random variable X that represents the number of different numbers that appear in* 38 *spins of the roulette wheel?*

If we choose a number k of distinct numbers to appear and ask what is $P(X = k)$, then we know that

$$P(X = k) = \binom{38}{k} p^k (1 - p)^{38-k}$$

where $p = 1/38$ and $q = (1 - p) = 37/38$. But as we saw earlier, such formulas can be difficult to compute by hand. It is here that our new distribution will help.

The idea behind the new approximating distribution is that $P(X = k)$ remains essentially the same when the number (k) and the number of plays (38 spins) are both enlarged by the same amount. Thus, if we consider a wheel with 100 numbers and we tested 100 spins, $P(X = k)$ would be about the same as with 38 spins.

To see why this is true, we examine the formula for $P(X = k)$. Here we assume the number of spins (trials of the experiment) is n and the probability of spinning k is p. Thus, $\lambda = np$ has a fixed value which in our case is $n(1/n) = 1$.

If we replace p with λ/n, then we see that

$$P(X = k) = \binom{n}{k} p^k (1 - p)^{n-k}$$

$$= \frac{1}{k!} \times \frac{n}{n} \times \frac{n-1}{n} \times \ldots \times \frac{n-k+1}{n}$$

$$\times \lambda^k (1 - \lambda/n)^n (1 - \lambda/n)^{-k}$$

$$\approx \frac{\lambda^k}{k! e^\lambda}$$

(This approximation comes from letting n grow large.)

We should view this probability distribution as a random variable Y (approximately X). The range of values for Y is $k = 0, 1, 2\ldots$ and the probability distribution (the *Poisson Distribution*) is given by

$$P(Y = k) = \frac{\lambda^k}{k! e^\lambda}.$$

In our example, with $\lambda = 1$, we obtain the values shown in Table 3.3, which are compared to the corresponding binomial distribution values.

We are finally ready to determine the expected number of distinct numbers obtained in 38 spins of the roulette wheel. To do this we define some new random variables

$$Y_0, Y_{00}, Y_1, Y_2, \ldots, Y_{36}.$$

where the random variable Y_i takes on the value 1 if i appears exactly once in 38 spins and 0 otherwise.

What is $EV(Y_0)$? From the definition of expected value and the probabilities from Table 3.3 we see that

$$EV(Y_0) = P(Y_0 = 0)(0) + P(Y_0 = 1)(1) = .36788.$$

TABLE 3.3:   Binomial and Poisson Values for $P(k)$, $0 \leq k \leq 8$

k	Poisson distribution $P(Y = k)$	Binomial distribution
0	.36787	.36285
1	.36788	.37293
2	.18394	.18647
3	.06131	.06043
4	.01533	.01427
5	.00307	.00262
6	.00051	.00039
7	.00007	.00005
8	.00001	.00000

Note that this computation applies to each of the random variables Y_i, $i = 0, 00, 1, 2, \ldots, 36$. Thus, by the linearity of expected value we have

$$EV(Y) = EV(Y_0) + EV(Y_{00}) + EV(Y_1) + EV(Y_2) + \ldots + EV(Y_{36})$$

and hence

$$EV(Y) = 38 \times .36788 = 13.979.$$

Thus, we conclude that the average number of numbers appearing exactly one time in 38 spins is about 14. We see from Table 3.3 that this computation is essentially the same for the number of numbers not appearing at all, that is,

$$EV(\text{number of numbers not appearing in 38 spins}) = 14.$$

Further note that $14/38 = .368 \approx 1/3$. Hence, approximately one-third of the numbers, on average, will not appear in the 38 spins of the roulette wheel and hence approximately two-thirds of the numbers will appear, with around 14 of these appearing only once!

What else can we say about the Poisson distribution?

Since the Poisson distribution is defined as

$$P(X = k) = \frac{\lambda^k}{k! \, e^\lambda} \tag{3.5}$$

we can determine the mean of the distribution as

$$EV(X) = \sum_0^\infty k \, P(X = k)$$

$$= \sum_1^\infty k \, \frac{\lambda^k}{k! e^\lambda}$$

$$= \lambda e^{-\lambda} \left(1 + \lambda + \frac{\lambda^2}{2!} + \dots \right).$$

But using a rule from calculus that

$$1 + m + \frac{m^2}{2!} + \frac{m^3}{3!} + \dots = e^m \tag{3.6}$$

we have that

$$EV(X) = \lambda e^{-\lambda}(e^\lambda) = \lambda.$$

A similar argument shows that $var(X) = \lambda$. Thus, the Poisson distribution has both mean and variance equal to λ.

After all this, some examples of how to use this distribution should be useful.

Example 3.4.1. *Suppose the week prior to a football game, fans arrive at the ticket window according to a Poisson distribution. Also suppose arrivals average 24 per hour. What is the probability of no arrivals in a given 5-minute period?*

Solution: The expected number of arrivals in a 5-minute period is

$$\frac{24}{60/5} = \frac{24}{12} = 2 = \lambda.$$

Thus, if X is the random variable counting the number of arrivals in the 5-minute window (using Equation (3.5)),

$$P(X = 0) = \frac{\lambda^0 e^{-\lambda}}{0!} = e^{-2} = .135.$$

□

Example 3.4.2. *A blackjack player has determined she averages a busted hand (with > 21 value) four times per hour. What is the probability she will have no busted hands in a 10-minute period?*

Solution: A 10-minute window is $1/6$ of an hour and so our player averages $\lambda = 4/6 = 2/3$ busted hands per 10-minute window. Thus,

$$P(X = 0) = \frac{(\lambda)^0 e^{-.667}}{0!} = .5132.$$

$\square$

Example 3.4.3. *A basketball player has a field goal percentage of 45%. With this shooting percentage, what is the probability the player makes eight straight shots? Now, in a season in which the player takes 800 shots, what is the probability the player has a streak of eight shots made in a row?*

Solution: First we must find the probability the player makes eight shots in a row. We use his shooting percentage as the probability he makes a shot. Thus,

$$P(\text{eight consecutive shots made}) = (.45)^8 = .0017.$$

Now that we have that probability, we can use the Poisson distribution to compute the probability he has such a streak. So,

$$\lambda = (800)(.0017) = 1.36.$$

Hence,

$$P(\text{eight shot streak happens}) = \frac{\lambda^1 e^{-1.36}}{1!} = .3490.$$

Thus, such a streak seems reasonably likely. $\square$

Exercises:

3.4.1 Records show that the probability a NASCAR race car will have a flat tire while driving one lap of a race is .00005. If during a race a total of $10,000$ laps are driven by the cars:

1. Find the probability exactly 1 will have a flat tire.

2. Find the probability at least 1 will have a flat tire.

3. Find the probability at most 2 will have a flat tire.

4. Find the probability at least 2 will have a flat tire.

3.4.2 If 2% of the baseballs manufactured are defective, find the probability a case of 300 baseballs contains

1. No defective baseballs.

2. Exactly 1 defective baseball.

3. Exactly 2 defective baseballs.

4. Exactly 3 defective baseballs.

5. At least 1 defective baseball.

3.4.3 A blackjack player has determined he receives a blackjack on 3% of the hands he is dealt. If he plays 200 hands, what is the probability

1. He will be dealt exactly 1 blackjack?

2. He will be dealt at most 2 blackjacks?

3. He will be dealt 4 or more blackjacks?

4. He will be dealt no blackjacks?

3.4.4 Out of 300 college football teams surveyed, the probability that their uniforms were one solid color was .02. In a sample of 50 college teams, find the probability that

1. Exactly 1 has solid color uniforms.

2. Exactly 2 have solid color uniforms.

3. None have solid color uniforms.

4. At least 1 team has solid color uniforms.

3.4.5 A baseball player has a lifetime batting average of .280. In a random sample of 40 at bats, determine the probability the player got

1. Exactly 15 hits.

2. Exactly 12 hits.

3. Exactly 11 hits.

4. Exactly 10 hits.

5. Exactly 13 hits.

3.4.6 A hockey player has a probability of receiving a penalty in any one game of .5.

1. What is the probability he receives no penalties in 10 consecutive games?

2. What is the probability he receives exactly 1 penalty in 10 consecutive games?

3. What is the probability he receives at most 2 penalties in 10 consecutive games?

3.5   Streaks — Are They Real?

Many people naively believe that when flipping a coin, if heads comes up, then tails should come up on the next flip, or at least be more likely to come up. Their misinterpretation of the laws of probability is based on the idea things are supposed to "even out." The laws of probability do imply this, but only for a large number of trials of the experiment in question, not immediately. Even then, we are not guaranteed that in 1000 flips of a fair coin there will be 500 heads and 500 tails. Rather, the number of heads and tails should both be near 500.

On the other hand, anyone who has ever gambled can tell you about "*streaks,*" both winning and losing. They remember them only too well. In this section we wish to examine the idea of streaks and see that they are really a natural part of random processes. In fact, if you play enough, your experience tells you streaks are to be expected.

Question 3.5.1. *What is the probability you lose five consecutive bets on red in roulette?*

Solution: The probability you lose on a red bet in roulette is .526. Hence, the probability you lose five consecutive bets is

$$P(\text{lose five consecutive bets}) = (.526)^5 = .0403.$$

Thus, there is around a 4% chance you will have such a losing streak. □

Although the last probability is not large, it is certainly positive and says that if you play enough you should expect such a streak to happen. We will look more closely at this idea later.

Example 3.5.1. *The probability you will lose any hand of blackjack is .52. What is the probability you will lose eight consecutive hands?*

Solution: Again, as the losing probability is fixed, this is a straightforward computation. Namely,

$$P(\text{eight consecutive losses}) = (.52)^8 = .0053.$$

Thus, there is about a .5% chance of such a streak. □

Example 3.5.2. *What is the probability you will have a winning streak of six consecutive bets on black at American roulette?*

Solution: We know that the probability of a win on black is .474. Thus,

$$P(\text{six consecutive wins}) = (.474)^6 = .0113.$$

Hence there is about a 1% chance of such a streak. □

Example 3.5.3. *A baseball player has a lifetime batting average of .310. What is the probability he will have no hits in 16 consecutive at bats?*

Solution: Since the player's success percentage is .310, that means his failure percentage is .690. Thus,

$$P(0 \text{ for 16 streak}) = (.690)^{16} = .00264.$$

□

Example 3.5.4. *What is the probability the same player will have 8 consecutive hits?*

Solution: This time we use the batting average of .310 as the probability of success. Hence,

$$P(8 \text{ consecutive hits}) = (.310)^8 = .0000853.$$

From this we see that the player is much more likely to have the 0 for 16 slump, then the hot streak of 8 for 8. □

The above examples show that there is a nonzero probability of many kinds of streaks. When the probabilities are nonzero, streaks can and usually do happen, provided we play enough. We next turn to measuring the likelihood of these things happening.

Example 3.5.5. *If our baseball player (with the .310 average) bats* 600 *times in a season, what is the probability he has a* 0 *for* 16 *slump?*

Solution: Here is a case where the Poisson distribution can help us. Let X be the random variable that counts the number of 0 hits in 16 at bat streaks in a season. We have a large number of trials (600 at bats) and a small probability of the 0 for 16 event. These are conditions when we can approximate this probability with the Poisson distribution. Here we have 585 different 16 consecutive at bat sequences, as every at bat except the last 15 is the first at bat of one such sequence. Thus, as $(585)(.00264) = 1.544 = \lambda$ we have that

$$P(\text{no such streak, i.e., } X = 0) = \frac{(1.544)^0 e^{-1.544}}{0!} = .2135.$$

So there is a probability of .2135 that no such slump will happen, meaning there is a probability of

$$1 - .2135 = .7865$$

that such a slump will happen! Such a high probability for a player with a high batting average says that 0 for 16 slumps are not really uncommon, but rather events to be expected, especially among players with a large number of at bats. $\qquad\square$

Example 3.5.6. *What is the probability that in* 200 *hands of blackjack, you will have an* 8-hand losing streak?

Solution: There are 193 such 8-hand sequences, all but the last 7 hands are the first hand of such a sequence. From Example 3.5.1 we know the probability of such a streak is .0053. Thus, if X counts the number of such streaks then

$$P(X = 0) = ((193)(.0053))^0 e^{-1.023}/0! = .3595.$$

Hence, the probability of at least one such losing streak is $1 - .3595 = .6405$. Hence, there is a fairly high probability of one or more such losing streaks. $\qquad\square$

Exercises:

3.5.1 If the Bulls have a probability of .55 of winning any game against the T-Wolves, what is the probability

    1. The Bulls win four consecutive games against the T-Wolves?

    2. The Bulls win six consecutive games against the T-Wolves?

    3. The T-Wolves win three consecutive games against the Bulls?

3.5.2 If Alex Rodriguez has a lifetime batting average of .306, what is the probability he

    1. Gets no hits in his next 8 at bats?

    2. He gets no hits in his next 12 at bats?

    3. He gets 5 hits in his next 5 at bats?

    4. During a season with 600 at bats, he has a 0 for 8 streak?

    5. During a season with 600 at bats, he has a 5 hits in 5 at bats streak?

3.5.3 If Lebron James has a career free-throw shooting percentage of .728, what is the probability

    1. He makes his next 4 free-throws?

    2. He misses his next 2 free-throws?

    3. He makes his next 10 consecutive free-throws?

    4. During a season where he shoots 800 free-throws, he has a streak of 10 consecutive free-throws made?

    5. During a season where he shoots 800 free-throws, he has a streak of 6 consecutive missed free-throws?

3.5.4 If the Yankees team's winning percentage is .560, what is the probability

    1. The Yankees lose their next 2 games?

    2. The Yankees win their next 6 games?

    3. The Yankees win their next 10 games?

    4. The Yankees lose their next 5 games?

    5. Over 100 games, they have a 7-game winning streak?

6. Over a 162 game season, they have a 7-game winning streak?

3.5.5 If Chipper Jones presently has a batting average of .375, what is the probability he will

1. Get a hit in each of his next 5 at bats?

2. Get no hits in his next 6 at bats?

3. Get no hits in his next 10 at bats?

4. During a season with 500 at bats, he will have a 0 for 10 batting streak?

3.6 Betting Strategies

Another aspect of gambling, as old as betting itself, is the idea of *betting strategies*. There will always be people who believe they can beat the house with some grand plan of how to bet. The first plan anyone thinks of is the classic "doubling strategy," known as the *martingale strategy*.

Originally, martingale referred to a class of betting strategies popular in France in the 18th century (see [24].) This strategy has the gambler double his bet after every loss and resume his initial bet after every win. By doubling the bet after every loss, the first win recovers all previous losses (in this losing streak), plus provides a profit equal to the value of the initial bet. Since the bettor expects to win eventually, the martingale system seems the perfect strategy, at least at first glance. However, there are several possible flaws in the system. To make the arithmetic easier, we first assume an initial bet of $1. Also, suppose the bettor loses his first seven bets. Then at this stage the bettor has lost

$$\$1 + \$2 + \$4 + \$8 + \$16 + \$32 + \$64 = \$127.$$

Note that this is a classic *geometric progression* (recall geometric series discussed earlier). That is, a finite sum of terms where the first term is some value a (here $a = 1$) and the ratio between consecutive terms is r (here $r = 2$). There is a well-known theorem on geometric progressions.

Theorem 3.6.1. *Given a sum*

$$a + ar + ar^2 + ar^3 + \ldots + ar^{n-1}$$

where r is a fixed number, then the sum of these values is given by

$$\frac{a - ar^n}{1 - r}.$$

Proof. Let S_n be the sum of the first n terms of this progression. Then, we see that

$$S_n = a + ar + ar^2 + ar^3 + \ldots + ar^{n-1} \qquad \text{and}$$

$$rS_n = \quad ar + ar^2 + ar^3 + \ldots + ar^{n-1} + ar^n.$$

Thus,

$$S_n - rS_n = a + 0 + 0 + \ldots + 0 - ar^n$$

hence,

$$S_n(1 - r) = a - ar^n$$

so that

$$S_n = \frac{a - ar^n}{1 - r}. \quad \square$$

Thus, playing under the martingale system (hence, $r = 2$), with an initial bet of \$$a$, and losing n consecutive bets means the next bet will be

$$a2^n.$$

Thus, in our example above, the amount of money lost would be $\$2^7 - 1$ and our next bet would be \$128 which would recover our lost \$127 and yield a \$1 profit if we win.

Let's now consider some examples.

Example 3.6.1. *Suppose your initial bet at blackjack was \$50 and you have lost 5 consecutive hands. Playing the martingale system, how much would you next bet?*

Solution: So far you have bet and lost

$$\$50 + \$100 + \$200 + \$400 + \$800 = \$50(1 - 2^5)/(1 - 2) = \$50(2^5 - 1) = \$1550.$$

Thus, under the martingale system our next bet should be \$1600. $\quad \square$

We note here that we already know that the probability of losing 5 consecutive hands at blackjack is $(.52)^5 = .038$, and hence we should expect nearly 4% of the time such a streak will happen.

Example 3.6.2. *Suppose your initial bet on red at American roulette was \$100 and also suppose you lost your first 6 spins. How much would you next bet under the martingale system?*

Solution: So far you would have lost

$$\$100 + \$200 + \$400 + \$800 + \$1600 + \$3200 = \$6300.$$

Thus, your next bet would be $100(2^6) = \$6400$.

We also note that a losing streak of 6 consecutive spins at roulette has a probability of $(.474)^6 = .0113$. Thus, this will happen on average about 1% of the time.   □

These last two examples help us see the weakness of the martingale system. First, by doubling your bets you are dramatically (in fact exponentially) increasing your wagers. As we saw in Example 3.6.1, under the martingale system you would be betting \$128 in an attempt to make a \$1 profit. Since you are still betting at a disadvantage, the house is happy to take this bet! Here is also where the house rules come into play. Every table game comes with a minimum amount you must bet and a maximum amount you are allowed to bet. The idea is that the maximum really makes the martingale strategy a dangerous one, as sometimes you will have a losing streak long enough that you can no longer double your bet. At that stage your strategy is worthless and you are also out of a lot of money!

Clearly, you can maximize the number of times you can double your bet by starting with the table minimum bet. But then, the amount you can win on any one of these attempts is also the table minimum. Hence, winning a large amount will be a slow and dangerous task.

Another betting system that is popular is the *anti-martingale* strategy. With this strategy the bettor actually believes in streaks and hopes for a winning streak. The idea is to *parlay* your bet, that is, combine your winnings from the last bet, with the initial bet, to make your next bet. Here the bettor is believing in hot streaks and that he is more likely to win on the next bet.

This system can produce rapid gains. Starting with just a \$10 initial bet, a winning streak of just 4 bets would leave the bettor with \$160, for a \$150 profit. The big problem is that a winning streak of 4 only happens with probability

$$(.474)^4 = .0505$$

at roulette, and with probability

$$(.48)^4 = .0531$$

at blackjack, so that most of the time, if you try for that fourth victory, you are just donating back some good winnings! Even worse, most people would not stop at the fourth victory.

Thus, clearly the key to success at the anti-martingale system is knowing when to quit and take your winnings. Actually, this is the secret of all gambling!

There are other somewhat mathematical systems that are common. One such system is called the *cancellation system*. In this system you decide how much you wish to win and then write down a list of positive integers whose sum equals the amount you decided upon. Note that the list can be as long or as short as you wish. The strategy is that on your next bet, you wager the sum of the first and last numbers currently on the list (unless of course there is only one number left on the list and you just bet that amount). If you win, you cross those two numbers off your list; if you lose, write the amount of the loss as a new number on the end of the list.

You continue betting until either you cross off all the numbers from your list, in which case you have won your desired amount of money, or you can't afford to bet any more. Now, as long as the number of losses do not outnumber the number of wins by a 2 to 1 margin or more, you will be crossing off more numbers than adding, and you should reach your target winnings as your list will shrink in length. As with the earlier systems, the flaw here is that your bet sizes may escalate, again reaching the house limit or your own financial limit.

Specific analysis of the cancellation system is difficult as the system of bets is so general and the sequence of outcomes random.

Problem 3.6.1. *You are rolling a pair of dice and you are playing the cancellation system with a list of*

$$3, 1, 2, 4, 5, 3, 2.$$

You will win if a 7, 3 or 11 is rolled and lose otherwise. Play the game to test the cancellation system.

The moral of the story in this section is: there are no systems for betting fixed probability games that are guaranteed to work. We know the expected value of any single bet will be negative and no fancy series of bets can change that fact. Even if you adopt some variable amount betting system, it cannot change the fact that each individual bet has a negative expected value. Thus, the casino is not worried about your system, an indication that perhaps you should not worry about it either.

Exercises:

3.6.1 Suppose instead of the standard martingale system you decided to play a tripling system, that is, after each loss you would triple your bet, until winning, at which point you restart with your initial bet.

1. Write a formula for the amount bet after n consecutive losses, if your initial bet is m dollars.

2. Provide a solution to the formula in part (1).

3. How much are you ahead if your first victory is on hand two?

4. How much are you ahead if your first victory is on hand three?

3.6.2 Repeat parts (3) and (4) from the previous question, but this time for the standard martingale system.

3.6.3 Suppose you play a cancellation system with a goal of $20. If your initial cancellation list is

$$3, 2, 4, 1, 2, 2, 1, 5$$

and your first five bets result in a loss, win, win, loss and loss, show the cancellation list after these five bets have been placed.

3.6.4 Suppose you play an anti-martingale system in an even money game with the following change: instead of parlaying your winnings, suppose you only parlay half your winnings on each bet. If your initial bet is $20, show the sequence of bets if you win three consecutive times.

3.6.5 Determine a formula for the amount that will be bet on the nth bet, using the anti-martingale system, with an initial bet of $1. Do the same with an initial bet of a dollars.

3.7   The Gambler's Ruin

The oldest dream of any gambler is to break the house, or more generally stated, to win all the money from his opponent so no further play is possible. In modern times the opponent is often the casino (hence, the reference to breaking the house).

Earlier we developed methods for computing the probability of being ahead (or behind) by some specified amount after a large number of plays of a fixed probability game. We used the standard normal approximation to estimate fairly accurately such probabilities. But now suppose we consider a more unlimited question:

Question 3.7.1. *What is the probability a gambler, with an unspecified number of bets, will break the bank before going broke himself?*

This is often called the *gambler's ruin* problem. In attempting to answer this question we will get yet another look at a gambler's long-term prospects against the house (or against any opponent who has a consistent advantage over him).

In order to attack this question, we first need to make some initial assumptions and designate some variables. Thus, suppose the gambler starts with m units of money, while the house has $t - m$ units (so there is a total of t units at stake). The gambler will make repeated 1-unit bets at even money in a fixed probability game until either he has all the money or the house has all the money. We let the fixed probability p denote the probability the gambler wins any one bet and so $q = 1 - p$ denotes the probability the house wins any given bet.

Clearly, the gambler's holdings will change, moving up or down one unit with each bet. This fluctuation will continue until either his holdings reach t (breaking the house) or his holdings reach 0 (the gambler's ruin!). The game will end in either of these two situations as one of the two sides is out of money.

For each $m = 0, 1, 2, \ldots, t$ let

$$p_m = P(\text{gambler succeeds given holdings of } m \text{ units}).$$

Of course

$$1 - p_m = P(\text{gambler is ruined given holdings of } m \text{ units}).$$

There is a clear starting value for p_0, that is

$$p_0 = 0.$$

Now we note that if $m = 1$, then the gambler has only one way to succeed, that is to win the next bet and then to succeed with holdings of $m = 2$ units. This fact can be expressed as

$$p_1 = p \times p_2.$$

Now, with holdings of $m = 2$ units, the gambler has two ways to succeed. He can win his next bet and then succeed with 3 units or he can lose his next bet and then succeed with 1 unit. We can express this fact as

$$p_2 = (p \times p_3) + (q \times p_1).$$

Finally, when the gambler has succeeded he has all t units of money and hence

$$p_t = 1.$$

Applying the same type of arguments that gave us the expression for p_2, we get the following set of equations for all $t + 1$ probabilities:

$$p_0 = 0 \quad \text{and} \quad p_t = 1$$

and now for $m = 1, 2, 3, \ldots, t - 1$

$$p_m = p \times p_{m+1} + q \times p_{m-1}, \tag{3.7}$$

$$\text{for } m = 1, 2, \ldots, t - 1.$$

The above system of equations is of a very special type and there exist powerful techniques to solve them. However, these techniques are beyond the scope of this book. Hence, we shall take a more elementary algebraic approach.

First we note that another way to write p_m is

$$p_m = p \times p_m + (1 - p) \times p_m = p \times p_m + q \times p_m.$$

Now, using this fact, Equation (3.7) can be rewritten as,

$$p \times p_{m+1} + q \times p_{m-1} = p \times p_m + q \times p_m$$

or

$$p \times p_{m+1} - p \times p_m = q \times p_m - q \times p_{m-1}$$

and hence,

$$p_{m+1} - p_m = \frac{q}{p}(p_m - p_{m-1}).$$

Expressing this equation for each value of $m = 1, 2, \ldots t - 1$ gives

$$
\begin{aligned}
p_2 - p_1 &= \tfrac{q}{p}p_1 \\
p_3 - p_2 &= \tfrac{q}{p}(p_2 - p_1) &&= p_1\left(\tfrac{q}{p}\right)^2 \\
p_4 - p_3 &= \tfrac{q}{p}(p_3 - p_2) &&= p_1\left(\tfrac{q}{p}\right)^3 \\
&\;\;\vdots \\
p_m - p_{m-1} &= \tfrac{q}{p}(p_{m-1} - p_{m-2}) &&= p_1\left(\tfrac{q}{p}\right)^{m-1} \\
&\;\;\vdots \\
p_{t-1} - p_{t-2} &= \tfrac{q}{p}(p_{t-2} - p_{t-3}) &&= p_1\left(\tfrac{q}{p}\right)^{t-2} \\
1 - p_{t-1} &= (p_{t-1} - p_{t-2}) &&= p_1\left(\tfrac{q}{p}\right)^{t-1}.
\end{aligned}
$$

If we add the left-hand sides of all these equations we see that there is a great deal of cancellation and we simply obtain

$$1 - p_1.$$

Adding the right-hand sides, factoring out p_1 and equating we obtain

$$1 - p_1 = p_1\left[\frac{q}{p} + \left(\frac{q}{p}\right)^2 + \left(\frac{q}{p}\right)^3 + \ldots + \left(\frac{q}{p}\right)^{t-1}\right].$$

Solving for p_1 we have

$$p_1 = \frac{1}{1 + \frac{q}{p} + \left(\frac{q}{p}\right)^2 + \ldots + \left(\frac{q}{p}\right)^{t-1}}.$$

Returning to our system of equations and repeating the above procedure on only the first $m - 1$ equations we obtain

$$p_m - p_1 = p_1\left[\frac{q}{p} + \left(\frac{q}{p}\right)^2 + \ldots + \left(\frac{q}{p}\right)^{m-1}\right].$$

Solving for p_m and substituting the value of p_1 from above we see that

$$p_m = \frac{1 + \frac{q}{p} + \left(\frac{q}{p}\right)^2 + \ldots + \left(\frac{q}{p}\right)^{m-1}}{1 + \frac{q}{p} + (\frac{q}{p})^2 + \ldots + \left(\frac{q}{p}\right)^{t-1}}, \tag{3.8}$$

for $m = 1, 2, \ldots, t-1$.

But now we note that we have obtained an old friend, a geometric progression, in both the numerator and denominator. Thus, we will be able to say even more. From this point on we break the analysis into two cases depending on the value of p.

Case 1: Suppose that $p = q = 1/2$.

In this case $p/q = 1$ and our equations reduce to

$$p_m = \frac{m}{t}, \quad m = 0, 1, 2, \ldots, t.$$

But now we can see that the gambler's probability of breaking the bank equals the ratio of his holdings to that of the total amount in play. But, even in this fair game, the holdings of one gambler are relatively insignificant in comparison to that of the house. Thus the probability the gambler wins is near zero from the outset of the game. This fact should discourage anyone with the idea of trying to break the house.

Case 2: Suppose $p \neq q$.

In this case, $q/p \neq 1$ and we can use geometric progressions to help us. Considering Theorem 3.6.1, we see that

$$1 + \frac{q}{p} + \left(\frac{q}{p}\right)^2 + \ldots + \left(\frac{q}{p}\right)^{k-1} = \frac{1 - (q/p)^k}{1 - (q/p)}$$

and applying this to our formula for p_m we obtain

$$p_m = \frac{1 - (q/p)^m}{1 - (q/p)^t} \text{ for } m = 0, 1, 2 \ldots, t.$$

Now to completely understand this general formula we must again consider two subcases.

Subcase 1: Suppose $q > p$ and k is large enough that $(q/p)^k$ is large compared to 1.

In this case p_m is approximately

$$\frac{(q/p)^m}{(q/p)^t} \approx \left(\frac{p}{q}\right)^{t-m}$$

that is, we can drop the 1 from both the numerator and denominator without really changing the outcome. But as t grows large with respect to m, this can be seen to grow close to zero as $0 < p < q < 1$.

Subcase 2: Suppose $p > q$ and t is large enough that $(q/p)^t$ is essentially equal to zero.

In this setting where the gambler has the edge, the house is almost certain to lose. But where can we ever find such a wonderful setting?

Hence, in all the real cases we are likely to find, the gambler is ruined!

Exercises:

3.7.1 Consider Subcase 1 in the case of American roulette where $p = .474$. Determine p/q and show the gambler's probability of breaking the house is approximately $(.9)^t$ and so even if the gambler and the house start evenly with 50 units, the gambler will be ruined with probability .005; that is, the gambler will be ruined 199 out of 200 times.

3.7.2 Do the same analysis of Subcase 1 for blackjack where we assume $p = .48$.

3.7.3 Do the same analysis as Exercise 3.7.1 for craps with $p = .493$.

Chapter 4

Card Tricks and More

4.1 Introduction

In this chapter we will consider several mathematically based card tricks and games, as well as some strange situations. The idea is that we will see uses for mathematics you might never have guessed existed. In order to learn these card tricks, we will extend our mathematical base by learning new counting techniques and some new math models. We will also make use of some of the tools we have already seen. What we shall see is that knowing a little mathematics can give you a big advantage in many situations!

4.2 The Five-Card Trick

In this section we will study an old two-person card trick that pretends to be a mind reading trick, when in reality it is simply an excellent example of encoding information. The trick was developed by Fitch Cheney in 1950 (see [29]) and is probably the best-known mathematical card trick. It incorporates several beautiful ideas and allows for interesting generalizations.

The Tools: The trick begins with a standard deck of 52 cards, you and a partner as well as a "victim" (who falls for the trick).

The Trick: You have the victim shuffle the deck as much as he wishes and then deal you five cards. Now you will lay four of these cards face-up on the table and one face-down. Your partner will be called in (have your partner nowhere close before this) and upon viewing the four face-up cards, your partner tells the victim the suit and rank of the face-down card. Your partner may want to put on a bit of a mind reading act while doing this!

The Mathematics Behind the Trick: This is not really a mind reading trick, but rather one of encoding information that your partner can read and use to determine the identity of the face-down card. Our first question is then:

Question 4.2.1. *What information do we need to encode?*

The face-down card carries two pieces of information, its rank and suit. These are the two things we must encode, using the four face-up cards. Then our partner will be able to read the information and determine the rank and suit of the face-down card. As an example, suppose after we place our cards face-up, they appear as below.

FIGURE 4.1:   The five-card trick.

The easiest bit of information is the suit. We are dealt five cards and thus, by *The Pigeon Hole Principle*, (see the statement below) we must have at least two cards of one suit.

Theorem 4.2.1. The Pigeon Hole Principle:
If $n + 1$ pigeons are placed into n pigeon holes, then one hole must contain at least two pigeons.

Proof: If each of the n pigeon holes contained only one pigeon, then there would be at most n pigeons. But we started with $n + 1$ pigeons, hence one hole has at least two pigeons.   □

The Pigeon Hole Principle sounds simple; however, it has many uses and powerful generalizations. At the very least, we should also be aware that there is a more general statement of the Pigeon Hole Principle.

Theorem 4.2.2. General Pigeon Hole Principle:

Let $q_1, q_2, \ldots, q_n$ be positive integers. If

$$q_1 + q_2 + \ldots + q_n - n + 1$$

pigeons are placed into n pigeon holes, either the first pigeon hole contains at least q_1 pigeons, or the second pigeon hole contains at least q_2 pigeons, ... , or the nth pigeon hole contains at least q_n pigeons.

Proof: The proof of this version works very much the same as the first version. If pigeon hole i contains at most $q_i - 1$ pigeons, then there are at most

$$(q_1 - 1) + (q_2 - 1) + \ldots + (q_n - 1) = q_1 + q_2 + \ldots q_n - n$$

pigeons in total. But, we have $q_1 + q_2 + \ldots + q_n - n + 1$ pigeons — a contradiction. □

Now, in our card trick, the suits are the pigeon holes, and hence we have four pigeon holes, but we have been dealt five cards. Thus we see we must have one suit with at least two cards.

We will use one of these cards as the face-down card and the other will be placed face-up in an agreed upon position. Without loss of generality, say the first card on the left in the row of four face-up cards is the card signaling the suit. Our partner then easily recognizes the suit of the face-down card. In the example of Figure 4.1 the suit of the face-down card must be clubs.

Next we must determine how to encode the rank of the card. We have three remaining cards that we will place face-up. These three cards can be anything else in the deck, so we must have a flexible rule for encoding the rank information. In order to do this we will take advantage of the fact there are 13 possible ranks. We picture the ranks as shown in Figure 4.2.

Note that any two of these values on the circle are separated by at most six from each other, along one side of the circle or the other! This separation of at most six is the crucial feature we will exploit!

For the three remaining cards, we wish to determine their relative ranks in order to have a low (L), medium (M) and high (H) card. To encode the information on the rank of the down card, we need the idea of a *lexicographic ordering*. By this we mean that one of these 3-tuples, say t_1, will appear before another, say t_2, provided in the first place where they disagree the entry of t_1 is less than that of t_2. Clearly, this

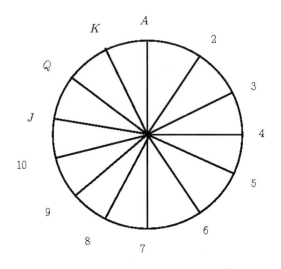

FIGURE 4.2:   The card rank circle (standard deck).

idea generalizes to k-tuples for any $k \geq 2$. For our 3-tuples, a natural ordering is $L < M < H$ and thus, the ordering of these three ranks produces the lexicographic ordering:

L M H (call this 1)
L H M (call this 2)
M L H (call this 3)
M H L (call this 4)
H L M (call this 5)
H M L (call this 6)

Note that any ordering can work, we just need to decide which one to use. However, the lexicographic ordering is easy to remember. Also note that all ties are broken based upon the alphabetic order of the suits of the cards. For example,

$$8\clubsuit < 8\diamondsuit < 8\heartsuit < 8\spadesuit.$$

Thus, with this ordering idea we can at least signal any value from one to six to our partner. What value we signal depends upon the two cards of the same suit in our hand. We will have the face-down card be somewhere between one and six "larger" than the face-up card (based on the rank circle traversed clockwise). Our partner will look at the first card, see the rank (as well as the suit) and determine the amount to add to that rank from the ordering of the next three cards, and thus determine the rank of the face-down card.

In our example, the first face-up card is the 8♣. The ordering of the other three cards is M, H, L which indicates we should add 4 to the 8 and hence signals a queen as the face-down card! Upon turning the card over we have the queen of clubs as seen in Figure 4.3.

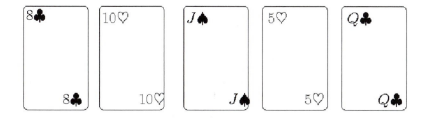

FIGURE 4.3:   The five-card trick completed.

Problem 4.2.1. *Suppose the cards you see are in order:*

$$7\heartsuit,\ 3\diamondsuit,\ 3\heartsuit,\ 10\clubsuit.$$

What is the face-down card?

4.2.1   Adding a Joker to the Deck

Now we add a joker to the deck and wonder if the trick can still be done.

Example 4.2.1. *Can we modify our techniques if we add a joker to the deck?*

Solution: If we are not dealt the joker, we just proceed as before. If we are dealt the joker, then there are two cases we must consider.

Case 1: We still have two cards in one suit.

In this case we place the joker face-down. We now do the trick as we normally would, placing the suit indicator card in place and ordering the other three cards to indicate the second card of that suit. But when our partner decodes this, she will also see that the indicated card is face-up among the three ordered cards. This is the signal that the joker is face-down.

Case 2: We are dealt the joker and one card of each of the other four suits.

In this case we pick a pair of cards that have their rank differences at most six. The lower (with respect to the rank circle) we put in the standard suit/base rank indicator position, the higher one becomes the face-down card. We now order the three remaining cards using the joker as the lowest (or highest — you decide which you prefer) of all cards. The fact the joker is face-up tells your partner you were dealt four suits and the joker, so the suit of the face-down card is the one missing from the table, not the one of the indicator card. The indicator card is just the base of the rank count and that value plus the value indicated by the three ordered cards tells your partner the rank of the face-down card.

There is one final situation to consider. Suppose you are dealt the joker and four cards of the same rank. In this case place the joker in the position of the indicator card and three of the rank cards face-up, the fourth face-down. The joker in the indicator position is the signal for this case and your partner will know the down card is the fourth card of the rank shown on the table.   □

Problem 4.2.2. *Suppose you see the following cards in order:*

$$5\clubsuit, \ Joker, \ K\spadesuit, \ 8\heartsuit$$

What is the face-down card in this situation?

4.2.2   More Variations of the Trick

Another natural question to ask is the following:

Question 4.2.2. *Can we further vary this trick to other sized hands and/or other sized decks?*

Everything fell into place for us in doing the five-card trick. The Pigeon Hole Principle worked perfectly for five cards of four possible suits. The 13 ranks on the wheel allowed us to have two cards whose distance in one direction or the other was at most six, and values from one to six were all we could communicate with permutations of three cards using lexicographic ordering. So we are in trouble when a $53rd$ card is added. The principles behind the trick break down and we must find other things that will allow us to perform the modified trick. We saw this with the joker, but it was special as the joker was not a part of any suit. Thus, there were ad hoc ways of handling this case.

In trying to answer Question 4.2.2 we should ask some other simple questions.

Question 4.2.3. *If I were dealt only four cards, what kind of deck would allow us to do a similar trick?*

Solution: In this case, with only four cards available to us, we must have only three suits in order to apply the Pigeon Hole Principle. With only three suits we are still guaranteed to have two cards in one suit.

Once we place one of these cards face-down, we are only left with three cards to encode information. One card must indicate the suit and base count for the rank of the face-down card. This leaves us with two cards to encode the displacement of the face-down card from the base rank card. But two cards can only convey $2! = 2$ possible permutations. Thus, our three suits may only have five cards (think of a five-card wheel, having distance at most two between any two of its five ranks, in one direction or the other). Thus, our deck would have to consist of only 15 cards, three suits with five cards in each suit. □

Example 4.2.2. *Is it possible to encode information differently and handle a larger deck? If so, what is the largest deck size d we can handle if we are dealt n cards?*

Solution: We have now turned this into a pure counting question. How many different signals can we send if we are dealt n cards? Again, being dealt n cards, we will place one card face-down and use $n - 1$ cards to transmit information. We will not be concerned with our own ability to remember all these possible signals as n grows large, but just whether we can create enough distinct signals. There are clearly $(n - 1)!$ ways to order these cards and as we have n choices for the face-down card, one hand of n cards can produce $n \times (n-1)! = n!$ different signals. Thus, we obtain an upper bound on the deck size as follows: We have $n!$ signals we can send and we can add the number of face-up cards to that total yielding

$$d = n! + n - 1$$

possible cards in our deck.   □

Our next question is a natural one.

Question 4.2.4. *Will we be able to handle the upper bound on d?*

As a test we consider a small case.

Example 4.2.3. *We consider the case when $n = 3$ with corresponding maximum deck size $d = 3! + 2 = 8$.*

Note that there are
$$8 \times 7 = 56$$

ordered pairs of eight numbers (let's think of each card as a distinct number for now). There are also

$$\binom{8}{3} = 56$$

unordered triples (the three-card hands) of eight numbers. Thus, for this case a possible strategy is a *bijection*, that is, a mapping that assigns to each of the unordered triples a unique ordered pair and each ordered pair appears exactly once.

More formally, given a mapping f from a set A to a set B, we call f a *bijection* if

(1) f is *one-to-one*, that is, for all $a, a' \in A$, with $a \neq a'$ we have $f(a) \neq f(a')$, and

(2) f is *onto*, that is, for each $b \in B$, $b = f(a)$ for some $a \in A$.

In our case A would be the set of unordered triples and B will be the set of ordered pairs. We seek a function f such that f assigns to each unordered triple a unique ordered pair (of elements from the triple). So in theory, such a mapping would yield a solution. The partner would see the two face-up cards in order and associate that ordered pair with one and only one triple of values. The value not seen is the face-down card. Note that such a mapping is possible when $n = 3$ and $d = 8$.   □

Question 4.2.5. *Can we always achieve the upper bound, that is, can we determine a bijection so that for each possible signal we can send, a distinct card can be determined?*

To answer this question we will employ a new model for the problem, a *graph*. A graph is an ordered pair (V, E) where V is a finite set of elements called *vertices* and E is a set of two element subsets of V called *edges*. This formal definition allows for a very visual model.

Example 4.2.4. *The drawing in Figure 4.4 shows how we might view the graph with vertex set*

$$V = \{v_1, v_2, v_3, v_4\}$$

and edge set

$$\{\{v_1, v_2\}, \{v_2, v_3\}, \{v_3, v_4\}, \{v_4, v_1\}, \{v_1, v_3\}\}.$$

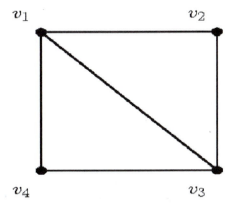

FIGURE 4.4: A drawing for the graph of Example 4.2.4.

Using graph models we can now attack the general upper bound problem. So let us assume a deck with $d = n! + n - 1$ cards and a hand of n cards. We are concerned with sets of size n (the hands) and ordered lists of length $n - 1$, that is, $(n - 1)$-tuples. Can we build a proper bijection to answer our question?

To attack this question, we will construct a graph model as follows: the vertices of the graph will be partitioned into two subsets. The first subset will consist of $\binom{d}{n}$ vertices, each representing one possible hand we could be dealt. The other set will contain $(n - 1)! \times \binom{d}{n-1}$ vertices, each representing a possible ordered list of $n - 1$ cards we could place face-up.

We now draw an edge between a "set" vertex and an "ordered list" vertex if and only if the elements of the ordered list are all in the set. The reader can now check that each of the two sets in our partition of the vertices has the same *cardinality*, that is, they contain the same number of vertices. Further, each vertex is connected by an edge to

$$\binom{n}{n-1} \times (n - 1)! = n! = d - (n - 1)$$

vertices of the other set. There are no other edges. Such a graph is called a *bipartite graph*, that is, a bipartite graph is one in which all edges are drawn between vertices in two disjoint sets, and these two sets partition the vertex set of the graph.

Now we would like to "pair" each set of n cards with one and only one ordered list and the elements of this list should be contained in the set. Luckily, there is a famous theorem that will help us. This theorem is known as **The Marriage Theorem** and it is due to P. Hall [17]. The goal of the theorem is that of pairing elements (like in a marriage). If we can successfully pair each element of one set of vertices of our graph with a distinct vertex of the other set and the two sets have the same cardinality, we obtain what is called a *perfect matching*. We state Hall's Theorem in a form that is easily applicable to our problem.

Theorem 4.2.3. Hall's Marriage Theorem
Suppose G is a bipartite graph with $V = A \cup B$ and $|A| = |B|$. Then the following are equivalent:
1. *G has a perfect matching.*
2. *Every subset of k vertices of A connects to at least k vertices of B.*
3. *Every subset of k vertices of B connects to at least k vertices of A.*

Now in our problem, each vertex has *degree* exactly $n!$, where the degree is the number of edges connecting a vertex to other vertices. We show by contradiction that a perfect matching must exist. Assume there is a subset of k vertices in A (or B works the same) that connects to fewer than k vertices of B. The number of edges incident to these vertices is exactly $k \times n!$. If there are fewer than k vertices in B incident to these edges then, by the general pigeon hole principle, one of the vertices of B must have more than $\frac{kn!}{k} = n!$ edges incident to it. But each vertex has only $n!$ edges incident to it. This contradiction shows that the conditions of Hall's Theorem must hold, and our graph has a perfect matching. This perfect matching determines the bijection we seek.

Example 4.2.5. *Consider a hand where $n = 2$ and hence the deck has $d = 3$ cards. Can we find the list that we will associate with each hand?*

Solution: Suppose for convenience that our deck is the A, K and Q of spades. There are three possible two-card hands we could be dealt, namely A, K or A, Q or K, Q. Our graph would appear as in Figure 4.5.

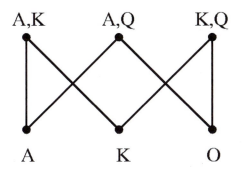

FIGURE 4.5: The bipartite graph obtained.

There are several ways to select a matching in this graph. For example the matching shown in Table 4.1.

From this table we see a matching and how to use it to do the trick in this case. For each possible two-card hand there is a distinct $n - 1 = 1$ card sequence that will be placed face-up and a unique card from the hand indicated by this sequence. This is the card that goes face-down. □

TABLE 4.1: A Matching in the Bipartite Graph

hand dealt	A, K	A, Q	K, Q
$n-1$ ordering matched to hand	A	Q	K
face-down card	K	A	Q

Exercises:

4.2.1 Suppose you are dealt $5\diamondsuit$, $10\heartsuit$, $2\heartsuit$, $A\clubsuit$, $A\spadesuit$ while doing the standard five-card trick. What is the face-down card and what is the order of the face-up cards (left to right)?

4.2.2 Suppose you are dealt $4\clubsuit$, $4\diamondsuit$, $4\spadesuit$, $6\heartsuit$ and the $7\heartsuit$. Determine the face-down card and the order of arrangement of the face-up cards.

4.2.3 You are doing the five-card trick with a joker included and you are dealt a hand with $A\spadesuit$, $6\spadesuit$, $8\heartsuit$, $7\clubsuit$ and the joker. What do you do?

4.2.4 You are doing the five-card trick with a joker included and you are dealt $4\clubsuit$, $5\diamondsuit$, $6\heartsuit$ and $7\spadesuit$ along with the joker. Now what can you do?

4.2.5 You are dealt 13 cards in a bridge hand. Then one suit must contain at least how many cards?

4.2.6 You are dealt 13 cards from a standard deck. Then one suit must contain at least how many cards?

4.2.7 If you wanted at least three clubs, or at least two diamonds, or at least five hearts, or at least four spades, how many cards must be dealt to you?

4.2.8 Determine a bijection that satisfies the $n = 3$, $d = 8$ case.

4.2.9 You are performing the five-card trick on a standard deck. Your partner has placed a card face-down and you see the arrangement in Figure 4.6. What is the face-down card?

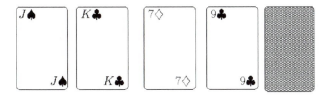

FIGURE 4.6:   You determine the face-down card.

4.3   The Two-Deck Matching Game

Next we consider a simple game, but one where many people guess incorrectly as to its likely outcome.

The Game: Begin with two standard decks of cards. Give one deck to person 1 and one deck to person 2. Each deck is shuffled as much as desired. Now person 1 and person 2 each turn over the top card of their decks. If the cards match exactly, the game is over. If not, they each turn over the next card of their deck. The game repeats until there is an exact match or the cards in the decks run out.

Question 4.3.1. *Would you bet even money that there will be an exact match before the deck runs out?*

When asked this question over the years, the majority of students have responded no. After some further questioning the following rough answer is sometimes given.

There is a $\frac{1}{52}$ chance the card in deck 2 will match that of deck 1. Since there are 52 tries at a match, the expected number of matches is $52 \times 1/52 = 1$. Thus, we should expect a match at some point in the game.

This rough argument allows one to believe a match is to be expected and hence a bet on a match occurring seems very reasonable. But can we say more here? Can we actually determine the probability there will be a match, thus giving us more information?

The solution to our question lies in a careful study of permutations. We have two standard decks of cards. We can view the second deck as a permutation of the cards in the first deck. Suppose there is an exact matching of cards and it occurs upon turning over the *ith* card of each deck. Since these cards are exactly the same, we are asking for the permutation representing the second deck to have a *fixed point*, that is, an element that did not change positions under the permutation.

There is no match only when deck 2 represents a permutation of deck 1 that has no fixed points. We call such a permutation a *derangement*. We can certainly count all the permutations of a standard deck of cards. There are 52! such permutations. Our goal is now to count those permutations that are derangements.

To count the derangements it is useful to actually count the permutations that are not derangements. To do this we will apply a well-known counting technique called the *inclusion-exclusion principle*.

Let $P_1, P_2, \ldots, P_m$ be properties of the objects in the set S we are considering. Let

$$A_i = \{x \in S \mid x \text{ has property } P_i\},$$

for $i = 1, 2, \ldots, m$, be the subsets of objects of S which have property P_i (and possibly other properties). Then $A_i \cap A_j$ is the subset of objects that have both properties P_i and P_j (and possibly others). Continuing along this line, $A_i \cap A_j \cap A_k$ is the set of objects that have properties P_i and P_j and P_k (and possibly others). The subset of objects having none of the properties is

$$\overline{A}_1 \cap \overline{A}_2 \ldots \cap \overline{A}_m.$$

The *Inclusion-Exclusion Principle* tells us how to count the number of objects having none of the properties, by counting those which do have properties.

Theorem 4.3.1. The Inclusion-Exclusion Principle: *The number of objects of a set S which have none of the properties $P_1, P_2, \ldots, P_m$ is given by*

$$|\overline{A}_1 \cap \overline{A}_2 \ldots \cap \overline{A}_m| = |S| - \sum |A_i| + \sum |A_i \cap A_j| - \sum |A_i \cap A_j \cap A_k|$$

$$+ \ldots + (-1)^m |A_1 \cap A_2 \cap \ldots \cap A_m|$$

where the first sum is taken over all $\binom{m}{1}$ of the 1-combinations of $\{1, 2, \ldots, m\}$, and the second sum is taken over all $\binom{m}{2}$ of the 2-combinations of $\{1, 2, \ldots, m\}$, and so forth.

Proof: The left-hand side of the equation counts the number of objects in the set S with none of the properties. We can establish the equation by showing that an object with none of the properties contributes a total of 1 to the right-hand side and an object with some of the properties contributes a total of 0 to the right-hand side.

So consider an object with none of the properties. Then since it is in S, and none of the other sets, its contribution to the right-hand side is

$$1 - 0 + 0 - 0 + \dots + (-1)^m \times 0 = 1.$$

Now consider an object x with exactly k of the properties ($k \geq 1$). Since x is in S, it contributes 1 to $|S|$. The contribution of x to $\sum |A_i|$ is k since it has exactly k of the properties and thus will appear in k of the sets A_i. The contribution of x to $\sum |A_i \cap A_j|$ will be exactly $\binom{k}{2}$ since x will occur in exactly $\binom{k}{2}$ of the sets $A_i \cap A_j$. In a similar manner the contribution of x to $\sum |A_i \cap A_j \cap A_k|$ is exactly $\binom{k}{3}$, and so forth. Thus, the contribution of x to the right-hand side of our equation is exactly

$$\binom{k}{0} - \binom{k}{1} + \binom{k}{2} - \binom{k}{3} + \dots + (-1)^m \binom{k}{m},$$

which equals

$$\binom{k}{0} - \binom{k}{1} + \binom{k}{2} - \binom{k}{3} + \dots + (-1)^k \binom{k}{k},$$

since $k \leq m$ and $\binom{r}{t} = 0$ if $t > r$. Since this last expression equals zero (use $(1-1)^k$ and the Binomial Theorem from Chapter 3), the net contribution of x to the right-hand side is zero and the theorem is proved. $\square$

In our case the set S is the set of all permutations of the standard deck of cards. The properties P_i are matching deck 1 in at least position i, for $i = 1, 2, \dots, 52$. Thus, a permutation with none of the properties is one with no matches to deck 1, that is, a derangement of deck 1.

Next we wish to determine the values of the other terms in the inclusion-exclusion principle. To do this we must count the number of permutations that will match deck 1 in at least position i (this places the permutation in set A_i). Consider a permutation where there is a match in position i. Then, in the other $n - 1 = 51$ positions of the permutation we may place any of the other $n - 1 = 51$ cards. There are clearly $(n - 1)! = 51!$ ways to do this. For those permutations with at least two matches to deck 1 we see that there are $n - 2 = 50$ other positions in which to place the $n - 2 = 50$ other cards. Hence, this can be done in $(n - 2)! = 50!$ ways. Continuing along this same line of reasoning, there must be $(52 - j)!$ ways to complete permutations with at least j matches to deck 1. Now, applying the inclusion-exclusion principle we see with $m = 52$ that

$$|\overline{A}_1 \cap \overline{A}_2 \ldots \cap \overline{A}_m| = |S| - \sum |A_i| + \sum |A_i \cap A_j| - \sum |A_i \cap A_j \cap A_k|$$

$$+ \ldots + (-1)^m |A_1 \cap A_2 \cap \ldots \cap A_m|$$

$$= 52! - \sum_{i=1}^{52} (51!) + \sum (50!) - \sum (49!) +$$

$$\ldots + (-1)^{52}(1!)$$

$$= 52! - \binom{52}{1} \times 51! + \binom{52}{2} \times 50! - \binom{52}{3} \times 49!$$

$$+ \ldots + (-1)^{52}(1!)$$

$$= \frac{52!}{0!} - \frac{52!}{1!} + \frac{52!}{2!} - \frac{52!}{3!} + \ldots + \frac{52!}{52!}$$

$$= 52! \left[\frac{1}{0!} - \frac{1}{1!} + \frac{1}{2!} - \frac{1}{3!} + \ldots + \frac{1}{52!} \right]$$

$$\approx 52! \times \frac{1}{e}$$

Note, the value $1/e$ is known from series approximation done in calculus. Thus, we see that the number of derangements is approximately $52! \times \frac{1}{e}$. That is, $1/e$ is approximately the fraction of all permutations that are derangements. But $1/e = .3679$. Thus, approximately 36.79% of the time there will be no match, while approximately $1 - .3679 = .6321$, or 63.21% of the time there will be at least one match. For an even money bet these are really rather good odds!

Exercises:

4.3.1 How many permutations are there for a set of four elements and how many of these are derangements?

4.3.2 Suppose you play the two-deck matching game with decks of size 4. What is the probability of at least one match?

4.3.3 Consider the previous question for decks of size 5.

4.3.4 Suppose you play the two-deck matching game for decks of 48 cards. Now what is the probability of a match?

4.3.5 In playing a certain card game you want each player to receive the same number of cards and all the cards of the deck must be dealt. If our game always has either five or seven players, how many deck sizes from 1 to 200 cards cannot be used in our game?

4.3.6 Do the same as in the previous problem for 3, 5 or 11 players and decks of sizes 1 to 1000.

4.3.7 A group of people wish to form teams of sizes 11 or 13. How many group sizes from 1 to 500 do not allow an even division of the players into teams of these sizes?

4.4 More Tricks

The next trick is based solely on arithmetic and is fairly easy to perform. We call it the *Prediction Trick*.

The Setup: Shuffle a standard deck of cards. Place the top card face-up on the table and based on its value (all face cards count 10, an ace counts as one, etc.) count out enough cards on top of this one (counting exactly one per each such card) to reach 10. (For example, if the face-up card was an 8, then two more cards would be placed on top of it to form the pile.) Now start a new pile with another face-up card. Keep making piles that reach 10 until the deck is used up. Note that there may be a few remaining cards that do not total 10. Just keep them.

The Trick: Ask some victim to choose three piles, each with at least three or more cards, and have them turn those piles face-down with the original card on top. Now pick up the remaining piles and place them, along with any left over cards in one pile.

Next you count off 19 cards from the leftover pile and place them aside. Have the victim turn over the top card (the one that was face-up) from any one pile and count off as many cards as that card's value from the leftover pile. Repeat this on a second pile.

Finally, count the cards remaining in the leftover pile and this will be the value of the top card on the third pile. If you can do the last counting inconspicuously, this will help sell the trick.

Example 4.4.1. *Why does this trick work?*

Solution: The trick is purely arithmetic. Each face-up pile starts with a card of value x and then $10 - x$ cards are added. Thus, there are $11 - x$ cards in that pile. When the three piles are considered, we see

$$(11 - x) + (11 - y) + (11 - z)$$

cards are in these three piles. Now we have removed 19 cards, then removed cards matching the value of the top card in two piles, without loss of generality, say these values are y and z. This accounts for

$$19 + (33 - x - y - z) + y + z$$

cards. That is, $52 - x$ cards have been dealt out, so the remaining pile must start with a card of value x. □

4.4.1 Friends Find Each Other

This is a very simple trick, but it looks nice when performed. We call it *friends find each other*!

The Deck: Have the victim place the ace, king, queen, jack and 10 of each suit in separate piles with the cards in that order in each pile. These cards will comprise the deck for this trick.

The Trick: Have the victim place any one pile on top of another and repeat until all four suits are in one pile. Offer the deck to the victim and ask him or her to cut it. Have someone else also cut the deck, perhaps several times. But make sure they are doing a single cut each time. After some fair number of cuts, begin dealing five piles of cards from the deck. Each pile will contain one rank. That is, the aces will all end up together, the kings all together, the queens all together and so forth.

The Solution: It is not to difficult to see why this card trick works. Concentrate on a single cut, say the first cut. What happens is the top x cards move to the bottom, but they remain in the same relative order. If you also think about the order of these cards to the other part of the deck, the cards remain in the same relative order viewed as a cyclic permutation. Thus, as the cards are dealt, since all the aces started out five apart, and they remain five apart, they will all land in the same pile.

4.4.2 The Small Arithmetic Trick

This is again a completely arithmetic-based trick and not as difficult to figure out as others we have seen.

The Trick: Have the victim shuffle a deck of cards that has had all 10s, jacks, queens and kings removed (aces serve as one here). Have the victim then draw one card from the deck without showing you the card. Then ask him or her to double the value of that card, then add 5 to the total. Now have him or her multiply that value by 5. Tell him or her to remember this number. Now have him or her draw a second card from the deck, again secretly. Have him or her add the value of that card to the total. Now have the victim announce the final value. You will then tell him or her the two cards her or she drew.

The Solution: The work for you is fairly easy. You subtract 25 from the total announced and the two digits of the value you get are the ranks of the two cards.

Why It Works: The arithmetic here is straightforward. Suppose the first card drawn had rank x and the second had rank y. Then the steps of the trick produce

$$((2x + 5) \times 5) + y$$

as the final value. That is

$$10x + 25 + y.$$

Once you subtract 25 from the total you have

$$10x + y$$

remaining, and since each of x and y is at most 9, we have a standard base 10 representation for our number and the first digit of the final value is x and the second digit is y. $\square$

4.4.3   The Nine-Card Trick

In this section we consider several fairly easy card tricks. Our problem for each of these is to determine why they work!

Deck: Have someone select any nine cards from a normal deck. Then let him or her shuffle these nine cards all they want.

The Trick: Ask the vicim to look at the third card from the top (make this seem random). Now, based on the name of the card he or she saw (for example the "three of clubs") have him or her deal a pile of cards, one by one, moving one card for each letter of each word. When any one word is complete, put the pile on the bottom of the nine card deck, then repeat the process for the next word. Make sure the piles are formed by top card from the deck to top of the pile! That is, they should spell out "three" (getting a pile of five cards that are then moved to the bottom of the original pile), then spell out "of" (then move that pile of two cards to the bottom of the original pile) and finally "clubs" (moving this pile of five cards to the bottom).

When the vicim is finished, you take the nine cards, look at the cards and tell the victim the card her or she saw, as it is always in the middle, that is, the fifth card from the top or bottom.

Added trick: Now do it again and have the victim lie to you, that is, he or she sees the third card from the top, but spells out any other card. It makes no difference, the middle card is still the one he or she saw.

Note, this only works for nine cards and the third card from the top. Why does this trick work? This is an exercise!

Exercises:

4.4.1 Explain why the nine-card trick works.

4.4.2 Explain why the nine-card trick works even if the victim lies to you about the card.

4.4.3 Suppose we make the following adjustment to the Prediction Trick. We count jacks as 11, queens as 12 and kings as 13. We deal as before, but we count up to 13 on each pile, rather than 10. After selecting the three piles, we discard 10 cards and perform the rest of the trick as before. Explain why this trick also predicts the top card of the last pile.

4.4.4 Can you find a different sequence of steps that allows you to do the Small Arithmetic Trick?

4.4.5 Will the Friends Find Each Other trick work on larger decks?

4.5 The Paintball Wars

In this section we will consider a game strategy for paintball when played by three players. Here the three players will have very different skill levels and this shall trigger the strategic keys. Suppose that we have three paintball players. Al (A) hits his target 50% of the time, Bob (B) hits his target 75% of the time and Carl (C) is deadly, hitting his target 100% of the time. In order to be more fair and take into consideration the different skill levels of the players, in the three-way paintball shoot-out that follows, each of the players shoots once in turn with the agreed upon order being Al, Bob, Carl, and then repeating as needed. Of course a dead player (hit by an earlier shot) loses his future turns. The major question for Al is what is his best strategy?

If Al kills Carl, then Bob will probably (75%) kill Al on his first shot. If Al kills Bob, then Carl will certainly kill Al (100%) on his first shot. Thus, things do not look good for Al. But there is one more option to consider, one that actually sounds a bit crazy at first, but turns out to really have merit. Al could intentionally miss on his first shot. This leaves Bob and Carl both still alive and both probably more worried about each other than about Al. Letting Bob and Carl shoot it out and then trying to shoot the winner of that battle may be a viable option. This is assuming Bob and Carl actually try to kill one another. But this assumption has merit. Bob will almost certainly choose to shoot at Carl rather than Al and if he survives, Carl would also almost certainly prefer to shoot at Bob, as Al is a much smaller threat.

How do we analyze the game strategy? It turns out that a tree diagram similar to what we did with the Penny Ante game will help. After Al purposely misses his shot, the general structure of the tree for the remainder of the game is shown in Figure 4.7. Note that missed shots create a right branch and made shots a left branch. The triangle labeled L represents the left subtree and will be examined in more detail after we examine the right subtree.

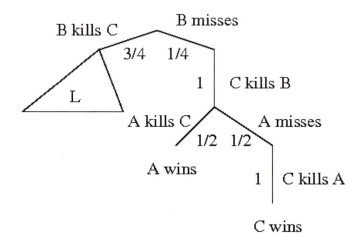

FIGURE 4.7:   The strategy tree diagram.

From Figure 4.7 we see exactly what happens in the right subtree. Following the paths downward, we see there are only two possible game endings, Al wins with probability

$$\frac{1}{4} \times 1 \times \frac{1}{2} = \frac{1}{8},$$

or Carl wins with probability

$$P(C \text{ wins}) = \frac{1}{4} \times 1 \times \frac{1}{2} \times 1 = \frac{1}{8}.$$

Note, this is the only case when Carl wins as he is killed by the first shot of the left subtree of outcomes.

Now consider the left subtree shown in more detail in Figure 4.8. From this tree we see that the situation is a bit more involved here. But, we can use the fact we are repeating A's turn in order to help simplify things. The subtree D is just this repeated pattern. Letting p be the probability that B wins from the given position in the tree where A is about to shoot (as shown with the arrows), we see that

$$p = \frac{3}{4} \times \frac{1}{2} + \frac{1}{2} \times \frac{1}{4} \times p.$$

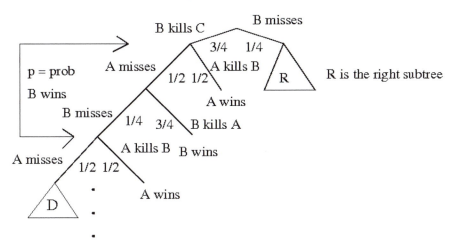

FIGURE 4.8: The left subtree.

Solving for p we see that $p = \frac{3}{7}$. Now the probability that B wins overall is

$$P(B \text{ wins}) = \frac{3}{4} \times p = \frac{9}{28} = .321.$$

Finally,

$$P(A \text{ wins}) = 1 - P(B \text{ wins}) - P(C \text{ wins}) = 1 - .321 - .125 = .554.$$

Thus, Al's strategy of missing on the first shot allows Al to have the best overall chance of winning the paintball game.

Exercises:

4.5.1 Suppose the game is played with probabilities 1/3, 2/3, 1. Now what are Al's chances of winning if he again plays the miss on the first shot strategy?

4.5.2 Suppose the probabilities are 1/2, 1 and 1. Now what are Al's chances of winning?

4.5.3 What happens in the game if Bob also purposely misses on his first shot?

Chapter 5

Dealing with Data

5.1 Introduction

The term statistics has the unfortunate fate of having two definitions that can cause a bit of confusion. First of all, *statistics* can be defined as a collection of numerical data. Most sports data fits into this definition and you often hear people refer to sports "stats" or to a player's "stats." This is how we think of sports data, and it is a proper view given this definition. Jay Bennett said, "Sports are statistics" and Leonard Kopett, a Hall of Fame sportswriter once said, "Statistics are the lifeblood of baseball" (see [4]). We will call such data *sports statistics*.

The second definition of *statistics* is the mathematics of the collection, organization and interpretation of numerical data. In this chapter we will begin such a study of statistics, but we will use sports statistics as our data. Our questions will include: How do we view large amounts of data? How do we measure this data? How do we compare this data with other data sets? What meanings can we derive from this data?

This is the real field of statistics, and one not usually applied to sports statistics. In fact, for many years no one took such a mathematical approach to sports data. However, in the 1950s, professional statisticians like Frederick Mosteller (in 1952), John Smith (in 1956) and Ernest Rubin (in 1958) became interested in applying statistical methods to baseball. But the leader during this period was George Lindsey who pioneered the use of statistical models in attempts to answer real questions about strategy and performance (see [4]). More recently, people like Jay Bennett, Jim Albert and Bill James have taken serious looks at various sports via statistics, especially baseball. Albert wrote one text [1] on teaching statistics using baseball, while Albert and Bennett wrote *Curve Ball*, a book about using statistics in analyzing baseball [2]. Bill James has written a number of books on baseball statistics and has certainly influenced the game with his work. However, we shall not be limited to baseball statistics in our study.

As stated earlier, our data will primarily be actual sports data. We will ask questions about this data, look at some common sports statistics and really see if they provide the answers we want about our sport, our favorite player or one particular game. If not, then maybe we can devise some other measures that better answer our questions.

We need to keep in mind that when we collect data we are not dealing with the entire set of outcomes. Statistics involves *sampling*, that is, obtaining a sample of the possible outcomes and making some decision or prediction based upon this sample. Thus, exact probabilities will not be known to us, but rather, approximations based upon the sample. Earlier we knew exact probabilities because we knew the sample space exactly. But for sports statistics, we only know what has happened so far. Our sample space is under construction and changing regularly.

In statistics, we use the term *population* to mean the entire set representing all outcomes of interest while a *sample* is a subset of measurements selected from the population of interest. We hope to use the sample to make predictions and generally to gain information about the population. Many times the entire population is simply too large to use as a data set or has yet to even be completed.

5.2   Batting Averages and Simpson's Paradox

Who is the best hitter in baseball? This is an age old argument among fans of the game. Whether it be present day hitters or all time greats, there seems to always be an argument over this question. In fact, entire books have been devoted to this argument. We now ask a simple question.

Question 5.2.1. *Can we measure hitting in such a way as to determine who is the best hitter in baseball?*

In trying to answer this question we should really ask the following question first:

Question 5.2.2. *How is hitting presently being measured? That is, what baseball statistic (or statistics) is used to measure hitting?*

The answer to this last question is not so simple. One might think of batting average as a measure of the best hitter. But every time I have asked this question of my students, they have not agreed that batting average is the measure of the best hitter. If batting average is not the best measure of hitting, is there any better measure? Here again we find no agreement. Generally, home run production, slugging and on-base percentage are not found to be better measures of hitters. When we try to use some or all of these statistics, the argument only seems to grow.

Thus, maybe we should step back from Question 5.2.1 and simply ask:

Question 5.2.3. *What does batting average actually measure?*

A player's *batting average* is defined as

$$\text{batting average} = \frac{\text{number of hits}}{\text{number of at bats}}.$$

It is the ratio of hits to at bats, almost like a probability of getting a hit, but not exactly that (as it is computed from a sample), even though we used it as an approximation of the probability earlier.

Looking at some examples, in the National League (on July 31, 2008), we see Chipper Jones with 117 hits in 317 at bats for a batting average of .369 and Albert Pujols with 120 hits in 338 at bats for a batting average of .355. While in the American League we see Alex Rodriguez with 106 hits in 323 at bats for a .328 batting average and Ian Kinsler with 146 hits in 447 at bats for a .327 batting average. If we said any one of these players was the best hitter in baseball, someone else would surely disagree.

Batting average measures just that, the fraction of times a player has gotten a hit. But batting is a combination of things including power (home runs), the ability to drive in runs, as well as just get hits. We are measuring just one thing with batting average, so it is not a big surprise there would be disputes over the question of who is the best hitter. Actually, it is our own interpretation of the word "best" here that causes the trouble, because if by best hitter I meant the player who got hits most often, then batting average would be the measure to answer the question.

But even then we need to be a bit careful. Batting averages are a proportion of the time the player gets a hit. As such we must be careful in how we interpret them. To see what we mean, consider the following example.

Question 5.2.4. *Consider the 2002 and 2003 batting averages of Ronnie Belliard and Casey Blake of the Cleveland Indians shown in Table 5.1. What can we see from this data?*

TABLE 5.1:   Comparison of Ronnie Belliard and Casey Blake

Year	Player	Hits	At Bats	Average
2002	Belliard	61	289	.211
2002	Blake	4	20	.200
2003	Belliard	124	447	.277
2003	Blake	143	557	.257
Totals	Belliard	185	736	.2514
Totals	Blake	147	577	.2548

The combined totals show Blake with a better overall batting average, even though Belliard had a higher average in each of the two years in question! What has just happened?

The example we have just shown is real. It demonstrates a common statistical situation called *Simpson's Paradox*. The idea is one sample's success rate (here batting average) for both data sets (here the two years) is higher than the success rate for the other sample. However, when the success rates for both data sets are combined, the sample with the lower success rate in each group ends up with the better overall success rate, thus the **paradox!**

When this happens, one sample usually has a considerably smaller number of members (here at bats) than the other sample. In our example we see Blake had considerably fewer at bats in 2002, so few that they had a small effect on his overall average. Simpson's Paradox does not occur in populations with similar amounts of data.

This phenomenon was described by Edward H. Simpson in 1951 and the name Simpson's Paradox was coined by Colin Blyth in 1972 [38]. However, it had earlier been described by others, including Udny Yule in 1903. Thus, some people call it the *Yule-Simpson Effect.*

We note that Simpson's paradox can work over longer periods than just two seasons.

Example 5.2.1. *Compare the batting averages of Derek Jeter and David Justice over the 1995, 1996 and 1997 seasons.*

Solution: If we build a table for these players over the three years we obtain Table 5.2:

TABLE 5.2: Comparison of Derek Jeter and David Justice: 1995–1997

Year	Player	Hits	At Bats	Average
1995	Derek Jeter	12	48	.250
1995	David Justice	104	411	.253
1996	Derek Jeter	183	582	.314
1996	David Justice	45	140	.321
1997	Derek Jeter	190	654	.291
1997	David Justice	163	495	.329
Totals	Derek Jeter	385	1284	.300
Totals	David Justice	312	1046	.298

When we look at the combined values for each player over the three years, we see that Derek Jeter has 385 hits in 1284 at bats for a .300 average while David Justice has 312 hits in 1046 at bats for a .298 average. Now after three seasons, each hitter has more than 1000 at bats and a comparison of their combined averages seems meaningful. But the relatively small number of at bats for Jeter in 1995 and Justice in 1996 help skew the yearly comparisons. □

Simpson's paradox is by no means rare. It has been found in data for many different types of questions and can cause some real problems in data interpretation.

Part of the problem here is that if $A_1 > B_1$ and $A_2 > B_2$ we are naturally inclined to believe that $A_1 + A_2 > B_1 + B_2$. But the fact is we are not adding these numbers (the batting averages), but rather adding the separate parts of the ratio to get a new ratio. Hence, the above inequalities do not apply.

TABLE 5.3:   Breakdown of Hitting Statistics for Two Players

	Player	Al		Player	Bob	
	Hits	At Bats	Ave.	Hits	At Bats	Ave.
Total	47	119	.395	45	119	.378
vs. L	30	65	.462	21	44	.477
vs. R	17	54	.315	24	75	.320
D	17	54	.315	24	75	.320
N	30	65	.462	21	44	.477
L-D	7	13	.538	11	21	.524
R-D	10	41	.244	13	54	.241
R-N	7	13	.538	11	21	.524
L-N	23	52	.442	10	23	.435

Example 5.2.2. *The data of Table 5.3 is taken from an example created by Allen Schwenk [37] . The batting totals for two players are shown. These totals are also broken down into several categories often considered in baseball. In the table, L= left-handed pitching, R= right-handed pitching, D= day game, and N= night game. In the exercises you will be asked questions about this data.*

Note that each category contains the complete totals for both players. All hits and at bats are shown for day or night games, or if the player hit against a left- or right-handed pitcher. The final set of numbers combines the two classes into four possibilities, facing a left hander during the day or at night and facing right handers during the day or at night.

In Chapter 6 we will look at some attempts to better understand offense in baseball. But most such attempts have not been widely accepted and hence, remain unknown statistics to the average fan.

We should also remark that judging hitting in baseball is not necessarily an easy task. In fact, as mentioned earlier, Michael J. Schell wrote an entire book [35] using statistical methods to try to decide who was the best hitter ever. His conclusion, after much work, was that Tony Gwynn was the best hitter ever, followed next by Ty Cobb.

Exercises:

Using the data from Example 5.2.2, answer the following questions:

5.2.1 Who had the better overall batting average?

5.2.2 Who had the better average against left-handed pitching? Who had the better average against right-handed pitching?

5.2.3 Who had the better average during day games? Who had the better average during night games?

5.2.4 Who had the better average against left-handed pitching during a day game? Who had the better average against right-handed pitching during day games?

5.2.5 Who had the better average against right-handed pitching during night games? Who had the better average against left-handed pitching during night games?

5.2.6 What conclusions can you draw from the example of Al and Bob?

5.2.7 Create your own example of Simpson's paradox.

5.3 NFL Passer Ratings

Although major league baseball has no formula for truly rating overall hitting (or overall offensive performance), other sports have been able to devise such ratings. The National Football League has accomplished what baseball has not, that is, they have produced a formula for rating quarterbacks. The idea is to have a numerical comparison based on a single number, rather than trying to compare all the various statistics a quarterback can produce. These statistics for quarterbacks include completions, pass attempts, touchdowns, interceptions, yards gained and more. Comparisons of all these values separately only opens the door for debate.

In 1973, retired NFL vice-president Don Smith conceived a formula that the league adopted for rating passers. The formula is based on four fundamental statistics, completion percentage, passing yardage, touchdowns, and interceptions. People could argue that this is too limited a scope, or the statistics are improperly weighted, but at least it is an attempt to quantify a quarterback's performance based upon recognized important statistics.

The rating is determined by four statistical components. Each of these is computed as a number between 0 and 2.375. The benchmarks for these values were based on historical data. The reader should be aware that the years immediately preceding 1973 marked a time in football where rushing the ball dominated over the passing game. Since then, rule changes have opened the game to much more passing. Those changes have produced much higher average quarterback ratings as the old benchmarks are low with respect to today's game.

The four components are (see for example [33]):

1. The component for completion percentage C is calculated as:

$$C = \frac{\left(\frac{\text{COMP}}{\text{ATT}} \times 100\right) - 30}{20}.$$

2. The component for yards per attempt Y is calculated as:

$$Y = \left(\frac{\text{YDS}}{\text{ATT}} - 3\right) \times \frac{1}{4}.$$

3. The component for touchdowns per attempt T is calculated as:

$$T = \frac{\text{TD}}{\text{ATT}} \times 20.$$

4. The component for interceptions per attempt I is calculated as:

$$I = 2.375 - \left(\frac{\text{INT}}{\text{ATT}} \times 25\right).$$

To help display the overall formula more simply, we now define some related parameters. For each of the four parameters, C, Y, T and I we define a corresponding parameter as

$$parameter' = max(min(parameter, 2.375), 0).$$

In particular then,

$$C' = max(min(C, 2.375), 0) \text{ and } T' = max(min(T, 2.375), 0).$$

The other two parameters are similarly defined.

The four new components are combined into one number for the rating as follows:

$$\text{QB Rating} = \frac{1}{6} \left[C' + Y' + T' + I'\right] \times 100.$$

The benchmark for completion percentage is 30 (the amount subtracted). Thus, to produce a positive value for C, the QB must have a completion percentage above 30%. Similarly, for yards per attempt the benchmark is 3. As stated earlier, these benchmarks would certainly be higher if the formula was modified for the present game.

Example 5.3.1. *Find the QB rating for Peyton Manning in 2006.*

Solution: The key statistics for Peyton Manning in 2006 were:

362 completions in 557 attempts for 4397 yards
with 31 TDs and 9 interceptions.

Based on these numbers we find:

- $C = ((362/557 \times 100) - 30)/20 = 1.7496.$

- $Y = (4397/557 - 3) \times \frac{1}{4} = 1.2235.$

- $T = 31/557 \times 20 = 1.2231.$

- $I = 2.375 - (9/557 \times 25) = 1.9711.$

We then see that $C' = C$, $Y' = Y$, $T' = T$ and $I' = I$. In fact, this will often be the case.

Finally, the QB rating for Peyton Manning in 2006 is:

$$\frac{1}{6} [1.7496 + 1.2235 + 1.1131 + 1.9711] \times 100 = 100.955 \approx 101. \quad \square$$

We next consider the career QB rating of a Hall of Fame quarterback, Dan Marino.

Example 5.3.2. *The career passing statistics for Dan Marino are: 4967 completions, 8358 attempts, for* $61,361$ *yards,* 420 *touchdowns and* 252 *interceptions. Determine his career quarterback rating.*

Solution: Calculating as before we find:

- $C = ((4967/8358 \times 100) - 30)/20 = 1.4714.$

- $Y = (61361/8358 - 3) \times \frac{1}{4} = 1.0854.$

- $T = 420/8358 \times 20 = 1.0050.$

- $I = 2.375 - (252/8358 \times 25) = 1.6212.$

Again, $C' = C$, $Y' = Y$, $T' = T$ and $I' = I$. Thus, the career QB rating for Dan Marino is:

$$\frac{1}{6} [1.4714 + 1.0854 + 1.0050 + 1.6212] \times 100 = 86.3833 = 86.4.$$

$\square$

The reader should note that the NCAA also has a passer rating formula. It is built (and correctly so) with different benchmarks. It also seems to be much less well known than its NFL counterpart.

Exercises:

5.3.1 Determine the maximum and minimum possible NFL QB rating.

5.3.2 John Elway had career statistics of 4123 completions in 7250 attempts for 51, 475 yards with 300 touchdowns and 226 interceptions. Compute Elway's career rating.

5.3.3 Compute the quarterback rating for one year on two quarterbacks of your choice.

5.3.4 If you were to modify the formula, how would you do it and why would your change be an improvement?

5.3.5 Test your modified formula against the standard one on three quarterbacks.

5.3.6 Determine what formula is used for the college (NCAA) quarterback rating. How is it different from the NFL formula?

5.3.7* In view of what you have seen on the quarterback rating formula, create a hitter rating formula for baseball.

5.3.8* Create an offensive player rating formula for basketball.

5.4 Viewing Data — Simple Graphs

Suppose we consider the number of home runs hit by each team in major league baseball in 2007. Table 5.4 shows these numbers. The data comes from [23]. What can we learn from this data set?

One way to begin to consider such data sets is to view the numbers in a *stemplot*. We note the home run totals vary from 231 hit by the Milwaukee Brewers to 102 hit by the Kansas City Royals. Our stemplot will reflect this by using 10–23 as the base numbers in the stem. Then for each base number we will record the third digit where it is appropriate. That is, if a team hit 171 home runs, we will record a 1 after the 17 in the stem. The stemplot for the 2007 season is shown in Table 5.5.

We note a unimodal (single local maximum) shape to the stemplot. This is information we would not have without "viewing" the data in

TABLE 5.4:　Home Run Totals for Both Leagues in 2007

American	League			National	League		
Team	HRs	Team	HRs	Team	HRs	Team	HRs
NYY	201	DET	177	PHI	213	COL	171
BOS	166	LAA	123	ATL	176	NYM	177
TEX	179	CLE	178	MIL	231	FLA	201
SEA	153	TBD	187	CIN	204	CHC	151
BAL	142	TOR	165	SDP	171	LAD	129
OAK	171	MIN	118	STL	141	PIT	148
KCR	102	CHW	190	HOU	167	ARI	171
				SFG	131	WSH	123

this manner. It is not unexpected that our sample data would have a shape roughly resembling the binomial distribution. We also note that more than half of the teams had home run totals between 141–179. This fact is easily seen from the stemplot.

Given a plot of some data set, we seek to gain all the information we can from this data set. To help in this, we make the following definition.

Definition 5.4.1. *Let $y_1, y_2, \ldots, y_n$ be a set of n measurements arranged in order of magnitude. The lower quartile (first quartile) is the value of y that exceeds 1/4 of the measurements and is less than the remaining 3/4 of the measurements. The second quartile is the median. The upper quartile (third quartile) is the value of y that exceeds 3/4 of the measurements and is less than the remaining 1/4 of the measurements.*

It is important to note that the quartile values may or may not be actual data points. It depends on the exact spread of the data as to how this might work. It is also important to note that you cannot always get exactly 25% of the data points to be less than the first quartile or greater than the third quartile. Sometimes you must settle for at least 25% of the values. This is because there may be many repetitions of values among the data.

Example 5.4.1. *Locate the first, second and third quartiles for the data plot of Table 5.5.*

TABLE 5.5: Stem Plot
of Major League HR Totals

10	2
11	8
12	339
13	1
14	128
15	13
16	567
17	111167789
18	7
19	0
20	114
21	3
22	
23	1

Solution: There are 30 ordered values in the stemplot of Table 5.5. Thus, $1/4$ of the values amounts to the first $\lceil 26/4 \rceil = 8$ values. We thus select a value greater than the 8th value (and less than the 9th value), namely 145. The median value will lie between the 15th and 16th values, which in this case are both 171, hence the median is 171. Finally, the third quartile must have at least 8 values greater than itself, hence we select 178.5. □

The quartile values are a help in measuring the spread of the data. We know 25% of the values are less than the first quartile (145) and 25% are greater than the third quartile (178.5).

Another measure concerned with the spread of the data is the sample variance s^2 and sample standard deviation s (see Section 1.9). If our sample X has n data values then

$$s^2(X) = \frac{\sum_{x_i \in X} (x_i - \mu)^2}{n - 1}$$

and

$$s(X) = \sqrt{\frac{\sum_{x_i \in X} (x_i - \mu)^2}{n - 1}}.$$

But we should keep in mind that for samples, an alternate way to compute the variance is

$$s^2(X) = \frac{n \sum x_i^2 - (\sum x_i)^2}{n(n-1)}.$$

Thus, we first determine the mean of the data set, which in this case is 165.2. We then compute the variance, which in this case is

$$s^2(X) = 890.788.$$

Finally, the standard deviation is the square root of the variance. Hence, in this case the standard deviation is 29.85.

We can exploit the idea of a stemplot to display information in other ways as well. The stemplot of Table 5.6 gives one such variation. Here we show the two leagues separately as well as the names of the teams achieving the particular totals. This gives us a different view of the information.

TABLE 5.6:   A Comparison Stem Plot of the Home Run Data

			KCR-2	10					
			MIN-8	11					
			LAA-3	12	WAS-3	LAD-9			
				13	SFG-1				
			BAL-2	14	STL-1	PIT-8			
			SEA-3	15	CHC-1				
		BOS-6	TOR-5	16	HOU-7				
TEX-9	CLE-8	DET-7	OAK-1	17	COL-1	SDP-1	ARI-1	ATL-6	NYM-7
			TBR-7	18					
			CHW-0	19					
			NYY-1	20	FLA-1	CIN-4			
				21	PHI-3				
				22					
				23	MIL-1				

Stem plots can vary in their shapes as well. It all depends upon the data at hand. Let's consider another example.

Example 5.4.2. *Draw a stemplot for the team batting averages for all major league baseball teams in the 2007 season. The data for this stemplot can be found in Table 5.7.*

TABLE 5.7: Team Batting Averages for 2007

	American	League			National	League	
Team	Ave.	Team	Ave.	Team	Ave.	Team	Ave.
TEX	.281	BOS	.281	CHC	.282	NYM	.267
CWS	.268	BAL	.272	STL	.281	PHI	.254
MIN	.278	DET	.274	MIL	.256	COL	.268
CLE	.258	NYY	.273	FLA	.250	ARI	.251
LAA	.265	TBR	.258	PIT	.261	ATL	.266
TOR	.262	SEA	.265	HOU	.263	CIN	.243
KCR	.261	OAK	.241	SDP	.248	SFG	.257
				LAD	.259	WAS	.245

Solution: The stemplot for team batting averages for the 2007 season is shown in Table 5.8. We again note a general unimodal shape to the stemplot. We also note that the majority of teams had batting averages between .250 and .268. The lower quartile is .2565, the median is .2625 and the third quartile is .2690.

Can you determine the mean, variance and standard deviation? The data comes from [23].

TABLE 5.8: Stem Plot for Team Batting Averages in 2007

24	1 3 5 8
25	0 1 4 6 7 8 8 9
26	1 1 2 3 5 5 6 7 8 8
27	2 3 4 8
28	1 1 1 2

□

We can use comparison stemplots to ask questions about data. For example, suppose we wish to compare league leading batting averages for the American League. Let's look at the decade of the 1920s and compare that to the decade of the 1980s. Here are the top averages for each decade.

American League leading batting averages in the two decades:

1920s: .407, .394, .420, .403, .378, .393, .378, .398, .379, .369
1980s: .390, .336, .332, .361, .343, .368, .357, .363, .366, .339

TABLE 5.9: Two-Decade Comparison
of American League Leading Averages

1920s		1980s
	33	6 2 9
	34	3
	35	7
9	36	1 3 8 9
9 8 8	37	
	38	
8 4 3	39	0
7 3	40	
	41	
0	42	

Now Table 5.9 clearly shows the 1920s produced higher league leading batting averages than the 1980s. This may be an indication of overall higher averages, but this is not a clear fact from this data. But what is clear is that in only two years of the 1980s was the league leading average as high as that of the lowest league leading average of the 1920s. The question then remains why there is such a visible difference? Of course one answer given to this is the change in the use of relief pitching in the modern game versus the game of the 1920s. But this is not determinable from the data plot. To prove this claim, other data and arguments would be needed.

What we can prove from this stemplot is that the 1920s saw American League leading averages that tended to be much larger than in the 1980s.

5.4.1 Time Plots and Regression Lines

There are other methods of displaying data that can also be useful. For example, consider the career data for a player like Cal Ripken. Ripken played from 1981 through 2001. In those years his home run totals are shown in Table 5.10 and the corresponding *time plot* of Figure 5.1. In the time plot, we show years along the x-axis and home runs along the y-axis. Because the points we are graphing are disconnected from one another, this type of graph is sometimes called a *scatter plot*. We will ignore the 1981 data, as he played very little that season.

TABLE 5.10: Cal Ripken's Home Run (HR) Totals by Year

Year	HRs	Year	HRs	Year	HRs
1981	0	1982	28	1983	27
1984	27	1985	26	1986	25
1987	27	1988	23	1989	21
1990	21	1991	34	1992	14
1993	24	1994	13	1995	17
1996	26	1997	17	1998	14
1999	18	2000	15	2001	14

If we plot a straight line through this data in a reasonable place, we see a general trend (see Figure 5.2). This trend shows that Ripken's home run production tended to decline as he aged. Of course, this is not an unusual event. But it is made clear to us by using the time plot and this "well placed" straight line. The idea is the straight line approximates the data.

But how do we find such a line. It is easy enough to draw in an approximation line modeling the data in this case, but what do we do

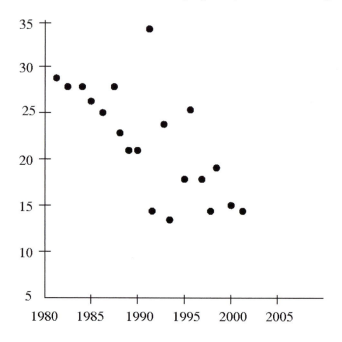

FIGURE 5.1:   Ripken time plot of home runs.

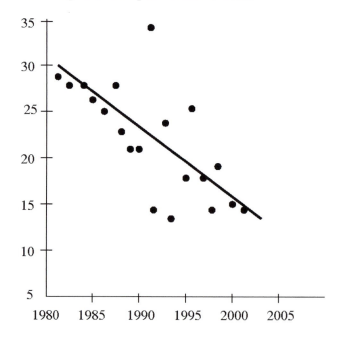

FIGURE 5.2:   Time plot with well-placed straight line.

in general to find the best line approximating the data?

Using the Ripken home run data of Table 5.10, suppose we consider a smaller subset of information. We will consider the years 1986 to 1989 only. This smaller example will be easier to describe. We now show Ripken's home run totals (see Table 5.11) and a time plot for this period of his career in Figure 5.3.

TABLE 5.11: Ripken Home Run Totals, 1986–1989

Year	HRs	Year	HRs
1986	25	1987	27
1988	23	1989	21

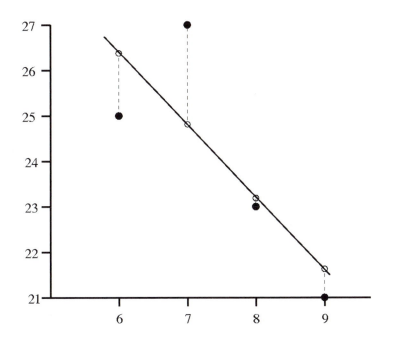

FIGURE 5.3: Ripken time plot for restricted period.

In the time plot, we have indicated the shortest distance from the actual data points to the line by a dashed line. The length of one of these lines can be thought of as the "error" in placement of the approximating line. The length of such a line can be found as the difference in y values of the endpoints of the lines. To find the best approximating line, we wish to minimize these errors taken as a whole. We cannot simply sum the errors to obtain the quantity we wish to minimize, as positive errors and negative errors would cancel one another, leading to false results. To eliminate this problem, we square the errors so that all values become positive. Thus, the sum of the squares of the error terms is the quantity we wish to minimize. The smaller this value, the better the approximating line.

We define the variable S_e to be the sum of the squares for the error terms in our time plot. If y is a data point and $\hat{y}$ represents the y value of the corresponding point on the line, then

$$S_e = \sum (y - \hat{y})^2.$$

We make the convention that the line that best fits the data is the line for which S_e is minimized. We call this line the *best fit line* or the *regression line* (sometimes called *the least squares regression line*).

Given a collection of n data points (x_i, y_i), $i = 1, 2, \ldots, n$, to find the best fit (regression) line we use the following method.

Let the equation of the regression line be defined as:

$$\hat{y} = b + mx,$$

where m is the slope of the line and b is the y-intercept (that is, the place where the line will cross the y-axis). This is of course the standard form for a linear equation.

Then we compute m as:

$$m = \frac{n(\sum_i^n x_i y_i) - (\sum_i^n x_i)(\sum_i^n y_i)}{n(\sum_i^n x_i^2) - (\sum_i^n x_i)^2}$$

and we define

$$b = \bar{y} - m\bar{x}.$$

Here $\bar{y}$ and $\bar{x}$ represent the average (mean) of the y and x values for the set of n data points.

In our example with data from Table 5.11:

$$\sum_i^n x_i = 30, \qquad \sum_i^n y_i = 96, \qquad \sum_i^n x_i y_i = 712$$
$$\sum_i^n x_i^2 = 230, \qquad \bar{y} = 24 \quad \text{and} \quad \bar{x} = 7.5.$$

Then,
$$m = \frac{4(712) - 30(96)}{4(230) - 900} = -1.6$$

and so
$$b = 24 - (-1.6)(7.5) = 36.$$

Thus, the equation for the regression line is then

$$\hat{y} = 36 - 1.6x.$$

Now we can measure the error terms for each data point. These values are:

$$e_1 = (25 - 26.4) = -1.4 \text{ and so } e_1^2 = 1.96$$
$$e_2 = (27 - 24.8) = 2.2 \text{ and so } e_2^2 = 4.84$$
$$e_3 = (23 - 23.2) = -0.2 \text{ and so } e_3^2 = .04$$
$$e_4 = (21 - 21.6) = -0.6 \text{ and so } e_4^2 = .36.$$

Hence,
$$S_e = 1.96 + 4.84 + .04 + .36 = 7.2.$$

Question 5.4.1. *Why would we want to find this regression line?*

The answer is simple, we wish to make predictions from this line. If our reasoning is correct and the line is a good approximation of the data, then we should be able to extend the line, thus making a prediction of data to come.

In particular, given our regression line $\hat{y} = 36 - 1.6x$, we can predict the home run total for Cal Ripken in 1990. Here we use $x = 10$ to follow the data used to construct the line, where we only used the final digit of the year in question to simplify the computations, that is, we use (Year - 1980) as the value.

Using $x = 10$ (to represent the year 1990) we obtain

$$\hat{y} = 36 - 1.6(10) = 20.$$

As you can tell from Table 5.10, Ripken's actual home run total for 1990 was 21. Thus, our prediction was very close to the truth.

Example 5.4.3. *Consider the following goal production of Wayne Gretsky: 1985–1986 he scored 52 goals, 1986–1987 he scored 62 goals, 1987–1988 he scored 40 goals and 1988–1989 he scored 54 goals. Do a time plot of Gretsky's goal production for this period and find the regression line. Then use the regression line to predict his goal scoring for the 1989–1990 season.*

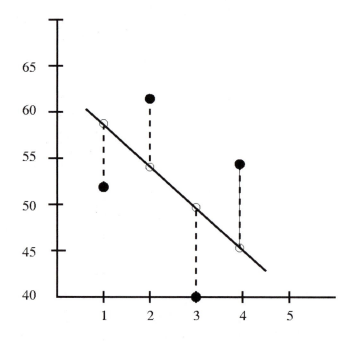

FIGURE 5.4:   Gretsky goal-scoring plot.

Solution: The plot of Gretsky's goal scoring is shown in Figure 5.4.

To compute the regression line, we follow the procedure from above, where again $n = 4$. For our example,

$$\sum_i^n x_i = 10, \ \sum_i^n x_i y_i = 512, \ \sum_i^n y = 208, \ \text{and} \ \sum_i^n x_i^2 = 30.$$

Further, $\bar{x} = 2.5$ and $\bar{y} = 52$.

Thus,

$$m = \frac{4(512) - 10(208)}{4(30) - 100} = \frac{-32}{20} = -1.6.$$

and $b = 52 - (-1.6)(2.5) = 56$.

Thus, the regression line is

$$\hat{y} = -1.6x + 56.$$

Now our prediction for the number of goals Gretsky would score in the next season (corresponding to $x = 5$) is $(-1.6)(5) + 56 = 48$. □

The real value for the goals he scored the next season is 40. This time the regression line is not as good a predictor as in the previous example. There can be many reasons for this, but one reason might be the limited number of points we used to build the line. Clearly, more information should be of help in such situations. The point is we are making a prediction based upon the sample and this prediction may be very good, or it may be far less good if we do not have enough information.

5.4.2 When To Find the Regression Line

Given a set of data points, we can always find the regression line by following our procedure. But we should ask if it makes sense to do this. That is, clearly sometimes a least squares regression line is not a good model for the data. For example, the points shown in Figure 5.5 clearly are not approximately linear. So any attempt to model this data with a straight line is clearly flawed.

We wish to quantify this almost linear relationship among the data points in some way. The closer the points are to the line, the stronger the degree of the linear relationship, called *linear correlation*.

We shall do this quantification by calculating a number r from the data. We will use r to help us determine if we even wish to find the regression line. This number r is called the *linear correlation coefficient*. We define r as

$$r = \frac{\sum(x - \bar{x})(y - \bar{y})}{(n - 1)s_x s_y}.$$

Here, s_x is the sample standard deviation of the x coordinates and s_y is the sample standard deviation of the y coordinates of the data points and $\bar{x}$ and $\bar{y}$ are the corresponding means for the x and y coordinates of the data points.

There is a somewhat easier equivalent computation for r, namely:

$$r = \frac{n(\sum_{i=1}^{n} x_i y_i) - (\sum_{i=1}^{n} x_i)(\sum_{i=1}^{n} y_i)}{\sqrt{n(\sum_{i=1}^{n} x_i^2) - (\sum_{i=1}^{n} x_i)^2} \times \sqrt{n(\sum_{i=1}^{n} y_i^2) - (\sum_{i=1}^{n} y_i)^2}}.$$

$$(5.1)$$

FIGURE 5.5: Data set that is not approximately linear.

We note, the closer that r is to $+1$, the better the points are to approximating a linear equation with positive slope, and the closer that r is to -1, the better the points are to approximating a linear equation with negative slope. An r value of 1 or -1 says the points are actually on the line. While an r value of 0 says there is no correlation between the points and a linear equation.

Let's apply this formula to the data sets in the examples where we found regression lines in the previous section.

Example 5.4.4. *Find the linear correlation coefficient for the home run data for Cal Ripken.*

Solution: In the example we computed the regression line for home runs hit by Cal Ripken, and made a fairly accurate prediction for the number of home runs he would hit the next year, based on the regression line. When we compute the linear correlation coefficient for that line, we obtain the following:

$$\sum xy = 712,\ \sum x = 30,\ \sum y = 96$$
$$\sum x^2 = 230,\ (\sum x)^2 = 900,\ \sum y^2 = 2324,\ \text{and } (\sum y)^2 = 9216.$$

Thus, the linear correlation coefficient is

$$r = \frac{4(712) - (30)(96)}{\sqrt{4(230) - (900)} \times \sqrt{4(2324) - (9216)}} \approx -.8004.$$

For the Ripken example, we obtained an $r = -.8004$ value. Since the regression line had negative slope, the $-.8004$ value says the points were reasonably well correlated to the line. □

Example 5.4.5. *Find the linear correlation coefficient for the data points of Example 5.4.3.*

Solution: Using the data from Example 5.4.3, we computed the regression line for Gretsky's goal scoring.

To compute the linear correlation coefficient r for these $n = 4$ data points, we again follow formula 5.1.

The key values were:

$$\sum_i^n x_i = 10, \ \sum x_i^2 = 30, \ \sum_i^n x_i y_i = 512,$$
$$\sum_i^n y_i = 208, \ \sum y_i^2 = 11064.$$

Hence, we see that

$$r = \frac{4(512) - (10)(208)}{\sqrt{4(30) - (10)^2} \times \sqrt{4(11064) - (208)^2}} = -.227.$$

Thus, we see the data points of the Gretsky example are far less correlated to the approximating line. In fact, the correlation coefficient is far closer to 0 than -1. This is an indication that the data points are not really well approximated by a linear equation. Thus, when our prediction was made, it is not a surprise it was not very close to the actual outcome. □

Our lesson here is that we should compute the linear correlation coefficient before we bother to find the regression line, so that we know ahead of time if it is worthwhile finding the line. We should also keep in mind that a strong correlation coefficient does not ensure that predictions made from the line will be accurate. It merely ensures that the data points are well correlated to a linear equation.

Exercises:

5.4.1 Table 5.12 shows the point production of two Hall of Fame NBA guards: Oscar Robertson and Jerry West. Do a comparative stemplot of this data.

TABLE 5.12: Comparison of Points Scored by Oscar Robertson and Jerry West

Year	Robertson	West	Year	Robertson	West
60–61	2165	1389	67–68	1896	1343
61–62	2432	2310	68–69	1955	1580
62–63	2264	1489	69–70	1748	2309
63–64	2480	2064	70–71	1569	1859
64–65	2279	2292	71–72	1114	1985
65–66	2378	2476	72–73	1130	1575
66–67	2412	1892	73–74	888	629

5.4.2 Given the stemplot from Exercise 5.4.1, what conclusions can you draw about this data?

5.4.3 Given the data of Exercise 5.4.1, find the first, second and third quartiles.

5.4.4 Using the data of Table 5.13, do a comparative stemplot for the National League batting leaders from the 1920s and the 1980s. How does this data set compare to that of the example given for the American League?

TABLE 5.13: Top National League Averages for the 1920s and 1980s

1920s	.370	.397	.401	.384	.424	.403	.353	.380	.387	.398
1980s	.324	.340	.331	.323	.351	.353	.334	.370	.313	.336

5.4.5 Given the data of Exercise 5.4.4, find the mean, variance and standard deviation the 1980s data.

5.4.6 Find the first, second and third quartiles for both sets of data from Exercise 5.4.4. What conclusions can you draw from this information?

5.4.7 Find the data and do a comparative stemplot for the league batting average leaders from the American League in the 1940s and 1990s. How does this stemplot compare to that of the 1920s and 1980s?

5.4.8 During the seasons from 1956 to 1960, Hank Aaron had 200, 198, 196, 223 and 172 hits. Do a time plot of this data.

5.4.9 Using the data of Exercise 5.4.8, find the regression line. Use this line to predict the number of hits Aaron would get in 1961. (Note: Aaron had 197 hits in 1961.)

5.4.10 During the five seasons from 1989–1990 to 1993–1994, John Stockton had 1134, 1164, 1126, 987, and 1031 assists. Do a time plot of this data.

5.4.11 Using the data of Exercise 5.4.9, find the regression line and use it to predict the number of assists Stockton had in his next season (note: he had 1011.) Find the linear correlation coefficient for this data.

5.4.12 During the six seasons from 1949–1950 through 1954–1955, Gordie Howe scored 35, 43, 47, 49, 33 and 29 goals. Do a time plot of his goal production for this period.

5.4.13 Given the data of the previous exercise, find the regression line and use it to predict the number of goals Howe scored the next season (note: he scored 38.) Find the linear correlation coefficient for this data.

5.4.14 Using the data of Exercise 5.4.12, find the mean, variance and standard deviation for Howe's goal production during this period.

5.5   Confidence in Our Estimates

Suppose you are the coach of a basketball team. You have a new player that you know very little about. In the first few games this player shoots 53% from the field making 53 of 100 shots. This seems excellent, but how sure are you that this reflects how the player will perform in the future?

Question 5.5.1. *How confident are we that this player is really a 53% shooter? Our question is really how confident are we that a sample of* 100 *shots accurately reflects this player's true shooting ability?*

Based on the Central Limit Theorem, we consider the number of shots made, x, to be binomially distributed. Given a player who makes x shots in n attempts, we can estimate (remember this is a sample) his true shooting percentage as

$$\hat{p} = \frac{x}{n},$$

his percentage for this sample. Since we assumed x is binomially distributed, the mean of this distribution is np and the standard deviation is $\sqrt{npq}$, where p is the true shooting percentage of the player and $q = 1 - p$. But we do not know the values of p or q, so we cannot compute either the mean or the standard deviation exactly.

Now recall our discussion of the meaning of the map (see Section 3.3)

$$Z = \frac{x - \mu}{\sigma}.$$

We know that Z is a measure of how many standard deviations x is away from μ. That is,

$$x - Z\sigma < \mu < x + Z\sigma.$$

We call the interval

$$[x - Z\sigma, x + Z\sigma]$$

a *confidence interval* for μ. That is, with some measure of confidence we feel the true value of the mean μ lies within this interval.

But how confident are we? Can we put a measure on our level of confidence? From a statistical point of view this would be ideal. Recall that Z gave us a probability from Table 9.1. Note that a value of $Z = 1.96$ produces an area to the left of 1.96 under the standard normal curve of .9750. This means, based on the symmetry of the standard normal curve, that the area to the right of 1.96 is .0250 and hence, the area to the left of -1.96 is also .0250. Thus, 95% of the area under the standard normal curve lies between -1.96 and 1.96. So if $Z = 1.96$, we would be 95% confident that our interval contained the true value. Other typically used Z values are shown in Table 5.14.

TABLE 5.14: Typical Z Values

Level of Confidence	Z Value
90%	1.65
95%	1.96
98%	2.33
99%	2.58

We now know how to compute a confidence interval for the mean of a distribution μ. But we were interested in our player's true shooting percentage p, not the mean μ for x. Luckily, we can make an adjustment for this proportion. We know the mean for the number of shots made is np, where n is the number of attempts and p is the true shooting percentage. Also, the standard deviation is $\sqrt{npq}$. Since x can vary from one sample to the next, so can $\hat{p}$. Thus, the sample probability $\hat{p}$ is approximately normal (when n is large enough, say $n \geq 30$), with mean

$$\hat{\mu} = \frac{np}{n} = p$$

and standard deviation

$$\hat{\sigma} = \frac{\sqrt{npq}}{n} = \sqrt{\frac{npq}{n^2}} = \sqrt{\frac{pq}{n}}.$$

It then follows that

$$\hat{Z} = \frac{\hat{p} - p}{\sqrt{\frac{pq}{n}}}.$$

Since we don't know p and q exactly, we will approximate them in the square root with $\hat{p}$ and $(1 - \hat{p}) = \hat{q}$, respectively. Thus,

$$\hat{p} - Z\sqrt{\frac{\hat{p}\hat{q}}{n}} < p < \hat{p} + Z\sqrt{\frac{\hat{p}\hat{q}}{n}}.$$

Now applying the same argument as earlier, a 95% confidence interval for the shooting percentage p would be

$$\hat{p} - (1.96)\sqrt{\frac{\hat{p}\hat{q}}{n}} < p < \hat{p} + (1.96)\sqrt{\frac{\hat{p}\hat{q}}{n}}. \tag{5.2}$$

Thus, for our problem, using $\hat{p} = .53$ and $\hat{q} = .47$ we obtain

$$.53 - (1.96)\sqrt{\frac{(.53)(.47)}{100}} < p < .53 + (1.96)\sqrt{\frac{(.53)(.47)}{100}}.$$

In other words, we have

$$.4322 < p < .6278$$

or, expressing it in interval notation, we have as our 95% confidence interval

$$[.4322, .6278].$$

We are 95% confident the true shooting percentage p for our player lies in this interval. But this interval has a fairly wide range, especially for shooting percentages.

Question 5.5.2. *What would be our 95% confidence interval if our player had taken 500 shots and made 265, for the same shooting percentage of .53?*

As with most things we have seen, more attempts (more data) give us more information. Recomputing Equation (5.2) with these new values gives

$$.53 - (1.96)\sqrt{\frac{(.53)(.47)}{500}} < p < .53 + (1.96)\sqrt{\frac{(.53)(.47)}{500}}.$$

Hence, we obtain a new 95% confidence interval of

$$[.4863, .5737].$$

Now we are 95% confident that our player has a true shooting percentage between .4863 and .5737. We have the same level of confidence over a much smaller interval. Thus, a larger data sample with the same level of success allows us to decrease the confidence interval's length, while maintaining the same level of confidence. The reader should note that if we wanted increased confidence but were using the original data set of 100 shots, we would have to increase the length of the interval.

Let's consider another example. Recall, in baseball, the *slugging percentage* is computed as follows:

$$\text{slugging percentage} = \frac{(1B) + (2 \times 2B) + (3 \times 3B) + (4 \times hrs)}{\text{at bats}}.$$

Example 5.5.1. *In 2006, Alex Rodriguez had 166 hits, including 26 doubles, 1 triple and 35 home runs in 572 at bats, for a slugging percentage of .523. Find a 99% confidence interval for A-Rod's true slugging percentage based on this sample.*

Solution: Slugging percentage is actually the mean number of bases per at bat. Confidence intervals for means are found using a similar formula where $\sqrt{pq/n}$ is replaced by $\sigma/\sqrt{n}$, where σ is the standard deviation. Here the standard deviation is computed on the number of bases per at bat and we find that approximately $\sigma = .851$.

Thus, a 99% confidence interval for A-Rod's true slugging percentage would be

$$.523 - (2.58)\frac{.851}{\sqrt{572}} < slug < .523 + (2.58)\frac{.851}{\sqrt{572}}.$$

Hence, we see that a 99% confidence interval for A-Rod's true slugging percentage is

$$[.4312, .6148].$$

□

Exercises:

5.5.1 In the first 30 games of a season, the Denver Nuggets had 1347 rebounds for a per game average of 44.87 and a standard deviation of 4. Find a 95% confidence interval for the Nuggets true per game average.

5.5.2 Suppose a pollster interviews 1000 NBA fans and finds 540 favor league expansion.

    1. Find an estimate for p, the true proportion of NBA fans favoring league expansion.

    2. Find a 95% confidence interval for p.

5.5.3 A basketball team sampled 75 fans to determine how many attend 5 or more games per season. Suppose 15 fans answered that they attend 5 or more games per season.

    1. Find an estimate for p, the proportion of fans who attend 5 or more games per season.

    2. Find a 90% confidence interval for p.

5.5.4 In a random sample of 120 football players, 20 admitted to using steroids.

    1. Estimate the proportion of players using steroids.

    2. Find a 98% confidence interval for the proportion of players using steroids.

5.5.5 If a sample of 50 baseball players from one league have a mean batting average of .260 with a standard deviation of .02, find a 95% confidence interval for the mean batting average in the league.

5.6   Measuring Differences in Performance

In this section we will consider several ways to compare performance of different players. We have already stated that the standard deviation is a measure of the spread of the data. We now present a theorem that explains how well standard deviation does this. The result is due to P.L. Tchebysheff (1821–1894).

Theorem 5.6.1. *Given a number $k \geq 1$ and a set of n measurements $y_1, y_2, \ldots, y_n$, at least $(1 - 1/k^2)$ of the measurements will lie within k standard deviations of their mean.*

We note that Tchebysheff's Theorem holds for any set of n measurements. We also note that it says at least a certain fraction of the data is contained within k standard deviations of their mean. It is well possible that even more of the measurements are contained in this interval, and this is the case for the normal distribution. We will say more about this later.

For $k = 2$, Theorem 5.6.1 says that

$$1 - \frac{1}{2^2} = 3/4$$

of the data lies within 2 standard deviations of the mean. Similarly, for $k = 3$ it says $8/9$ of the data lies within 3 standard deviations of the mean. The larger k is, the larger the fraction of data points guaranteed to be within k standard deviations of the mean. Note that for $k = 1$ the theorem provides no information about the distribution of the data.

Example 5.6.1. *In* 2007, *Curtis Granderson had* 185 *hits. There were* 126 *players in the American League with* 300 *or more at bats (see Table 5.15). Based on these players, how would Granderson rank?*

Solution: We restricted our attention to the American League players with 300 or more at bats. These were the players that played most regularly. We wish to rank Granderson against all other regular players. Now we find that the mean for this data distribution is

$$\bar{h} = 133.865,$$

the sample variance is

$$s^2 = 1375.03,$$

and the sample standard deviation is

$$s = 37.0814.$$

Since Granderson had 185 hits, while the mean was 133.865 hits, we see that $185 - 133.865 = 51.135$ and $51.135/37.08 = 1.379$. Thus, Granderson was 1.379 standard deviations above the mean. Hence, applying Tchebysheff's Theorem, we see that Granderson ranked above at least

$$1 - \frac{1}{1.379^2} = 1 - \frac{1}{1.902} = 1 - .5258 = .4742 = 47.42\%$$

TABLE 5.15:   Hits for American League Players with 300 At Bats in 2007

Suzuki	238	Ordonez	216	Jeter	206
Young	201	Polanco	200	Cabrera	192
Lowell	191	Rios	191	Markakis	191
Cano	189	Guerrero	186	Young	186
Granderson	185	Crawford	184	Rodriguez	183
Ortiz	182	Roberts	180	Hill	177
Sizemore	174	Guillen	172	Vidro	172
Hunter	172	Posada	171	Abreu	171
Martinez	169	Stewart	167	Ibanez	167
Guillen	167	Pedroia	165	Beltre	164
Ellis	161	Morneau	160	Blake	159
DeJesus	157	Matsui	156	Peralta	155
Teahen	155	Betancourt	155	Huff	154
Tejada	152	Youkilis	152	Cuddyer	151
Harris	149	Cabrera	149	Thomas	147
Figgins	146	Lofton	145	Hafner	145
Damon	144	Ramirez	143	Wells	143
Konerko	142	Upton	142	Rodriguez	141
Crisp	141	Swisher	141	Garko	140
Iwamura	140	Johjima	139	Pena	138
Grudzielanek	137	Pena Jr.	136	Lugo	135
Bartlett	135	Casey	134	Gordon	134
Lopez	132	Sheffield	131	Kotchman	131
Matthews Jr.	130	Dye	129	Mora	128
Byrd	127	Kinsler	127	Drew	126
Willits	126	Anderson	124	Pierzynski	124
Patterson	124	Millar	121	Inge	120
Uribe	120	Thome	119	Mauer	119
Kubel	114	Payton	111	Varitek	111
Kendrick	109	Castillo	106	Sosa	104
Wigginton	104	Stairs	103	Overbay	102
Barfield	102	Glaus	101	Cust	101
Punto	99	Johnson	98	Izturis	97
Butler	96	Owens	95	Brown	94
Hernandez	94	Gload	92	German	92
Laird	91	Fields	91	Sexson	89
Scutaro	88	Navarro	88	Tyneri	87
Catalanotto	86	Piazza	85	Gomes	85
Chavez	82	McDonald	82	Iguchi	82
Zaun	80	Wilkerson	79	Crosby	79
Erstad	77	Nixon	77	Buck	77
Monroe	76	Cruz	72	Vazquez	69

of the regular players. This means he ranked higher than $(126)(.4742) \approx 60$ of the regular players. We can easily see from the table that he actually ranked higher than many more players than that, in fact he ranked higher than 113 of the regular players.

We also see from the table that Sean Casey had 134 hits, so he was essentially on the mean. Also, Jorge Posada had 171 hits, so he was about 1 standard deviation above the mean and Macier Izturis had 97 hits, so he was about 1 standard deviation below the mean. Also,

Derek Jeter had 206 hits and was about 2 standard deviations above the mean. By Tchebysheff's Theorem we should expect that

$$1 - \frac{1}{(2)^2} = \frac{3}{4}$$

of the regular players were ranked below Jeter. That again is easily seen to be true, as he actually ranked above all but two other players that year. □

Example 5.6.2. *Suppose at a swim meet, 50 swimmers had 50-meter times in the range of 48 seconds to 66 seconds with a mean time of 54 seconds and a standard deviation of 2 seconds. How many of the swimmers had a time of between 50 and 58 seconds?*

Solution: The range 50–58 represents 2 standard deviations from the mean. Thus, by Tchebysheff's Theorem there are at least

$$\left(1 - \frac{1}{k^2}\right) = 1 - \frac{1}{4} = \frac{3}{4}$$

of the swimmers in this range. Thus there are at least

$$\frac{3}{4} \times 50 = 37.5 \approx 38$$

swimmers in this range. □

Example 5.6.3. *Suppose that the mean goals scored in a season for a hockey league of 100 players is 20 with a standard deviation of 8. If a player scored 40 goals, then this player outscored at least how many other players?*

Solution: Scoring 40 goals in this league means the player was 2.5 standard deviations above the mean. Applying Tchebysheff's Theorem, the player outscored at least

$$1 - \frac{1}{(2.5)^2} = 1 - \frac{1}{6.25} = .84$$

or 84% of the players in the league. Thus, he outscored at least $100 \times .84 = 84$ other players. □

We end this section with an empirical rule that applies to distributions that are roughly binomial (that is, "mound shaped"). This provides stronger fractions than we find in Tchebysheff's Theorem, since these rules apply to a smaller number of distributions.

- Approximately 68% of the data items will fall within 1 standard deviation of the mean (compared with no information from Tchebysheff's Theorem).

- Approximately 95% of the data items will fall within 2 standard deviations of the mean (compared with at least 75% from Tchebysheff's Theorem).

- Essentially all the data items will fall within 3 standard deviations of the mean (compared with at least 8/9 of the data items in Tchebysheff's Theorem).

5.6.1   Coefficient of Variation

The standard deviation of a data set clearly depends on the unit of measurement for that data set. For example, if the data is the weights of a collection of objects, the standard deviation may be something like 1 ounce. But this value tells us nothing about whether this reflects a huge deviation or a small deviation. If the objects are tiny, this standard deviation may actually reflect huge variations, while if the objects weigh hundreds of pounds each, this standard deviation reflects almost no variation.

We can normalize things by considering the *coefficient of variation* V. Given a data set $x_1, x_2, \ldots, x_n$ with mean $\bar{x}$, and standard deviation σ, we define the coefficient of variation as:

$$V = \frac{\sigma}{\bar{x}}.$$

The coefficient of variation expresses the standard deviation σ as a percentage of the mean of what is being measured and gives us another way to compare two different distributions. We now demonstrate this with an example.

Example 5.6.4. *During the past few months of workouts, one runner, Al, averaged 12 miles per week with a standard deviation of 2 miles, while another runner, Bob, averaged 25 miles per week with a standard deviation of 3 miles. Which of the two runners was relatively more consistent in their running workouts?*

Solution: Here we will rely on the coefficients of variation for the two runners. For Al,

$$V(\text{Al}) = \frac{2}{12} = .167$$

while for Bob

$$V(\text{Bob}) = \frac{3}{25} = .120.$$

Thus, we see that Bob was relatively more consistent in his workouts, even though his standard deviation was larger. □

Example 5.6.5. *At a tennis tournament, Pete Sampras had 4, 6, 2, 5, 1, 3 and 2 double faults in his matches, while Andre Agassi had 5, 2, 6, 4, 6, 4, and 3 double faults. Which player was more consistent in the sense of number of double faults per match?*

Solution: For Pete Sampras we find that his double fault data has mean and standard deviation

$$\mu_S = \frac{23}{7} = 3.2857,$$

$$\sigma_S = 2\sqrt{\frac{19.7634}{6}} = 3.2939,$$

and hence, a coefficient of variance of

$$V_S = \frac{3.2939}{3.2857} = 1.0025.$$

While for Andre Agassi, we see that his double fault data has mean and standard deviation

$$\mu_A = \frac{30}{7} = 4.2857,$$

$$\sigma_A = \sqrt{\frac{19.0608}{6}} = 3.1768,$$

and hence, a coefficient of variance of

$$V_A = \frac{3.1768}{4.2857} = .7413.$$

Thus, we see that Agassi was more consistent in his double fault pattern. □

Example 5.6.6. *Let us again consider the two sets of American League leading batting averages over the decades of 1920s and the 1980s. We ask which decade was more consistent in its leading batting average?*

Solution: For the decade of the 1920s we find a mean league leading batting average of .3897 with a standard deviation of .0196. Thus, the coefficient of variance is .0504, or 5.04%.

For the decade of the 1980s we find the mean leading batting average was .3375 with a standard deviation of .0168 and hence a coefficient of variance of .0497 or 4.97%.

Thus, the decade of the 1980s produced slightly more consistent league leading batting averages than the decade of the 1920s.   □

5.6.2   Relative Performance

In this section we consider another method of comparison of performance, developed by T. Oliver [31] in 1944. This concerns performance relative to the rest of your team. We begin with an example.

Example 5.6.7. *In 2008, Gil Meche had a win-loss record of 14–11 for the Kansas City Royals, whose overall record for that season was 75–87. Can we quantify how valuable Meche was to the Royals that season, relative to the rest of the pitchers on his team?*

Solution: We begin by noting that Meche's win-loss percentage for 2008 was $14/25 = .560$. Removing Meche's record from the team record, Kansas City was 61–76 for a win-loss percentage of $61/137 = .445$ in games in which Meche was not involved in the decision.

We now consider the effect of the difference in percentages between Meche and the team without Meche, over the games that Meche won or lost. We see that

$$\left(\frac{14}{25} - \frac{61}{137}\right) \times 25 = (.560 - .445) \times 25 = 2.875.$$

How do we interpret this computation? We see that this means that the Royals won 2.8 more games (because of Meche) than they would have won with an average performance by other pitchers from their team. Thus, Meche brought approximately an extra 3 wins to the Royals in 2008 than would otherwise have been the case.   □

Example 5.6.8. *Cliff Lee was 22–3 for the Cleveland Indians in* 2008. *The Indians finished the season with an 81–81 record. How valuable was Lee relative to the other pitchers on the Indians in 2008?*

Solution: We proceed as in the last example. Lee's win-loss percentage in 2008 was $22/25 = .880$. The Indians pitching staff without Lee was 59–78 for a win-loss percentage of $59/137 = .431$.

Thus, Cliff Lee's relative worth to the Indians in 2008 was

$$\left(\frac{22}{25} - \frac{59}{137}\right) \times 25 = (.880 - .431) \times 25 = 11.23.$$

Hence, we see that Cliff Lee provided an extraordinary 11–12 more victories than an average Indians pitcher would have provided. □

Clearly, such performances relative to "average performance" by the rest of the team can be computed for many other individual player statistics besides win-loss percentage for pitchers. It should also be clear that not all examples will provide "extra" value to the team. Here is an example from hockey.

Example 5.6.9. *In the 2007–2008 NHL season, Nikolai Khabibulin of the New Jersey Devils was 23–20–6 (wins-losses-overtime losses). The other goalies on the team combined for a 17–14–2 record. How valuable was Khabibulin relative to the other goalies on the team?*

Solution: Here we will consider overtime losses as losses. Thus, Khabibulin had a win-loss percentage of $23/49 = .469$. The remaining goalies had a win-loss percentage of $17/33 = .515$.

Thus, relative to the other goalies on the team, Khabibulin's relative worth to New jersey was

$$\left(\frac{23}{49} - \frac{17}{33}\right) \times 49 = -2.254.$$

This indicates that New Jersey lost about three more games than it would have with average performances from the other goalies. □

Exercises:

5.6.1 One NFL running back's yardage, for games over one season, averaged 102 yards per game. If the league average for starting backs was 72 yards per game, with a standard deviation of 12.5 yards, how did our running back compare to the rest of the league if there are a total of 32 teams?

5.6.2 One basketball league's records show that the average number of points scored by a player was 8.3 per game, with a standard deviation of 3 points per game. At least what percentage of the players in the league scored between 2.3 and 14.3 points per game? Using Tchebysheff's Theorem, at least what percentage of players scored at most 17.3 points per game? Suppose instead we assume a binomial distribution. Then, using the empirical bounds, what percentage of the players scored between 2.3 and 14.3 points per game?

5.6.3 A quarterback and an offensive lineman spend the off-season working out together. They are weighed each week. The quarterback has a mean weight of 200 over this period, with a standard deviation of 3 pounds. The lineman has a mean weight of 330 over this period, with a standard deviation of 5 pounds. Which player did a better job of maintaining his weight?

5.6.4 Two runners are dieting to reduce their weight. The first belongs to an age group with an average weight of 146 pounds and a standard deviation of 14 pounds. The second runner belongs to a group with average weight 160 pounds and a standard deviation of 17 pounds. If the two runners have weights of 178 and 193 pounds, respectively, which runner is more overweight?

5.6.5 Two hitters on different teams are worried about their batting averages. The first hitter plays for a team with a team batting average of .270 and a standard deviation of .015. The second hitter plays for a team with an team batting average of .260 and a standard deviation of .01. The first hitter is batting .250 and the second hitter is batting .245. Which hitter is weaker with respect to his own team?

5.6.6 If Rodger Federer has a first-serve percentage of 58% with a standard deviation of 3%, at least what percentage of matches will his first-serve percentage lie between 50% and 66%?

5.6.7 If Rodger Federer has a first-serve percentage of 58% with a standard deviation of 3% and Rafael Nadal has a first-serve percentage of 56% with a standard deviation of 2.5%, who is the more consistent with his first-serve?

5.6.8 Using the data from the previous problem, at least what fraction of the time will Federer have a first-serve percentage between 49% and 67%?

5.6.9 At least what fraction of the time will Nadal have a first-serve percentage of between 51% and 61%?

5.6.10 If men's pro tennis players as a group have an average first-serve percentage of 49% with a standard deviation of 3%, where does Rodger Federer (58%) lie with respect to this group? That is, at least what fraction of the players have a first serve percentage below his? What about Nadal?

5.6.11 In 2008, C. C. Sabathia had a win-loss record of 11–2 for the Milwaukee Brewers, who finished with a 90–72 record. How valuable was Sabathia relative to the rest of the team?

5.6.12 In the 2007–2008 season, Cam Ward of the Carolina Hurricanes had a goal tending record of 37–30 while the other goalies had a combined record of 6–8. How valuable was Ward compared with the other goalies on his team?

Chapter 6

Testing and Relationships

6.1 Introduction

In this chapter we take a look at typical statistical tests to verify claims and to test for the relationship between certain quantities. We want techniques for settling some common sports arguments or for proving some old sports adages actually hold. We also want to enhance the skills gained in the previous chapter so that we can use these skills to help answer questions of interest.

6.2 Suzuki Versus Pujols

In this section we will use some of the techniques we have learned to attempt to answer the question: In terms of batting average, which of Ichiro Suzuki of the Seattle Mariners or Albert Pujols of the St. Louis Cardinals is the better hitter?

This question is typical of many arguments among baseball fans or sports fans in general. Both players have had tremendous success, with a sustained period of high batting averages. Fans will argue broadly about which player they believe is the better hitter. We want to gather data and compare these players using statistical tests we have developed. We wish to see if our tests can supply the answers that fan arguments never seem to settle. Along the way, we must keep in mind that observed data does not represent the actual values, but rather reflects the variances that can occur randomly.

In some sense this is not a fair comparison. Ichiro Suzuki had a number of years of experience playing in Japan's top league before he came to the U.S. in 2001. Albert Pujols also began in the major leagues in 2001, but he was certainly younger and less experienced than Suzuki. Another thing that is somewhat unfair in the comparison is that they play in different leagues and thus they face different pitchers, in different stadiums with different playing conditions. (Except during interleague play where they will face some of the pitchers the other player faces. But the majority of each player's at bats are against his own league.) Note that we are assuming each player has a single true batting average.

We shall ignore these possible problems and concentrate on the data for each player in the period 2001–2008. Both players have shown excellent skills, batting above .300 each season. Albert Pujols' career batting average is .334 (1531/4578) while Ichiro Suzuki's career batting average is .331 (1805/5460). Although Pujols' average is slightly higher, Suzuki has nearly 300 more hits. These are the kind of facts that lead to fan arguments.

TABLE 6.1:   Pujols Versus Suzuki, Batting Average Stemplot

Pujols		Suzuki
	.30	3
4	.31	02
97	.32	12
110	.33	
	.34	
97	.35	01
	.36	
	.37	2

Table 6.1 shows a comparison stemplot of their yearly averages over this period. Pujols' averages have almost exclusively ranged between .329 and .359, with only one year at .314. Suzuki's averages have covered a much wider range. From that standpoint, Pujols' performance appears to be more consistent.

Next we take a look at a year by year plot of their averages (see Figure 6.1). Pujols' averages are shown by the solid curve and Suzuki's averages by the dashed curve. The time plot again confirms the more volatile changes in Suzuki's batting averages year to year, especially in the period 2003–2005.

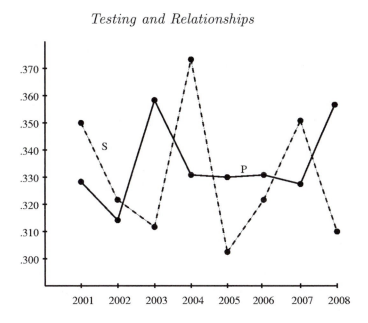

FIGURE 6.1: A time plot of batting averages for Pujols and Suzuki.

Since each player has more than 4500 at bats, we next compute confidence intervals for their true batting ability over this period, using their career numbers as the observed averages. We seek a 95% confidence interval on their true batting ability.

First we consider Suzuki. A 95% confidence interval for his true batting ability would be:

$$.331 - (1.96)\sqrt{\frac{(.331)(.669)}{5460}} < ave < .331 + (1.96)\sqrt{\frac{(.331)(.669)}{5460}}.$$

Hence, we obtain the interval [.319, .344] for Suzuki. That is, we are 95% confident Suzuki's true batting ability lies in this region.

Turning to Pujols, our computation of a 95% confidence interval for his true batting ability is:

$$.334 - (1.96)\sqrt{\frac{(.334)(.666)}{4578}} < ave < .334 + (1.96)\sqrt{\frac{(.334)(.666)}{4578}}.$$

Hence, for Pujols we obtain [.320, .348] as the 95% confidence interval of his true batting ability.

As we see, the two confidence intervals greatly intersect. From this it is impossible to claim which player actually has the higher ability. What we have been able to conclude is that Pujols has been the more consistent hitter year to year. But we can make no justified claim as to who is the better hitter. Let this argument rage on!

But despite our failure here, such tests can certainly work in other cases and for other measures besides batting average. We tried a variety of methods, but in this case we could reach only limited statistical conclusions.

Exercises:

6.2.1 Select two major league baseball players of your own and run a comparison of their batting averages. Attempt to determine which player has the higher actual batting ability.
(1) Repeat this test for their yearly runs batted in (RBI) totals.
(2) Repeat this test for their yearly home run totals.

6.2.2 Select two NBA basketball players and run a comparison of their yearly scoring averages in an attempt to determine which player has the higher scoring ability.
(1) Repeat this test for their yearly free throw percentages.
(2) Repeat this test for their yearly rebounding totals.

6.2.3 Select two hockey players and run a comparison of the yearly scoring for each player. Attempt to determine which player has the higher scoring ability.
(1) Repeat this test on their yearly assist totals.

Note: Data for these questions can easily be found on a number of different Web pages including at espn.go.com.

6.3   I'll Decide If I Believe That

Suppose a football coach states that the average college football player in the United States weighs 250 pounds. Should we believe him? That is, should we accept this statement as valid? The average person would accept this statement without giving it much thought.

But we have the tools to test this statement, at least to within some probabilistic range.

This common statistical test is called a *hypothesis test*. There are several variations of hypothesis tests and we shall consider some of these variations. Our example above amounts to testing a hypothesis about the mean of a population, namely the average weight of a college football player.

In testing claims, there are two fundamental hypotheses involved. The *null hypothesis* which is a statement asserting no change, no difference or no effect. Such a statement usually takes the form of a statement about a population parameter (like the mean) equaling some value. We label the null hypothesis with $\mathcal{H}_0$.

The typical hypothesis test also involves a second hypothesis, called an *alternative hypothesis*, and denoted $\mathcal{H}_a$. This is a statement that might be true instead of the null hypothesis. The alternative hypothesis usually involves parameters being greater or smaller than particular values, or possibly not equal to some value.

Thus, in our present example we have the following:

$\mathcal{H}_0$: the mean weight of a college football player is 250 pounds.
$\mathcal{H}_a$: the mean weight of a college football player is not 250 pounds.

The idea in hypothesis testing is to give the null hypothesis the benefit of the doubt. Just as a person is presumed innocent and must be proven guilty, the null hypothesis is presumed initially to be true and it must be shown, beyond some reasonable (probabilistic) level of doubt, that the null hypothesis should be rejected.

Suppose a random sample X of 100 college football players produced a mean weight of $\bar{x} = 246$ pounds, with a standard deviation of $s_{\bar{x}} = 20$ pounds. We will use this sample to test the hypothesis.

As we indicated earlier, the process for testing the null hypothesis is to first assume it should be accepted. We then wish to determine what values of $\bar{x}$ are so far from 250 pounds that such values would be very unlikely to occur if 250 pounds was really the mean weight of the population of all college football players.

The method used to test which $\bar{x}$ values are likely and which are unlikely is to use the probability distribution of $\bar{x}$. Recall, the Central Limit Theorem tells us that when $\bar{x}$ is obtained from a random sample of size n and n is large enough (and $n \geq 30$ works), then $\bar{x}$ will be approximately normal. In addition, the mean $\mu_{\bar{x}} = \mu$ and the standard error of the mean is $s/\sqrt{n}$. That is, the sample mean should be (approximately) the population mean and the standard deviation of the sample mean (standard error) should be related to the population standard deviation as shown.

For our sample of 100 college football players' weights we have $\mu_{\bar{x}} = 246$ pounds and $\sigma_{\bar{x}} = 20$ pounds.

Now, when we say that values of $\bar{x}$ are so far from 250 pounds as to be unlikely, we mean that such values have a very low probability of occurring. It is always up to the person performing the test to determine how low this unacceptable probability will be. However, it is common to select a value of something like .05, .02, or .01 but other values can be used. We call this selected probability the *significance level* of the test and denote it by α.

It is also more convenient to work with the standard Z value for $\bar{x}$, rather than $\bar{x}$ itself. That is, we again take advantage of the standard normal curve (as we have a large sample).

For our purposes then,

$$Z = \frac{\bar{x} - \mu_{\bar{x}}}{\sigma_{\bar{x}}} = \frac{\bar{x} - \mu_{\bar{x}}}{\sigma/\sqrt{n}}.$$

But we do not always know σ, as often the entire distribution is too large to obtain this value. When this is the case we may substitute the value of $\sigma_{\bar{x}}$ as a good approximation.

In our example, we obtain

$$Z = \frac{246 - 250}{20/10} = -2.000.$$

This is called the *test value* of Z. Recall that Z measures the number of standard deviations by which the value of $\bar{x}$ is above or below the mean. A positive value indicates that $\bar{x}$ is above the mean and a negative value indicates that $\bar{x}$ is below the mean.

Also recall that 1.96 was the $|Z|$ corresponding to a probability of .95. Thus, an observed z value with $Z < -1.96$ or $Z > 1.96$ falls outside the area we would expect. This area is called the *critical region*. That is, the region under the standard normal curve with $z < -1.96$ or $z > 1.96$ forms the critical region. Values of Z in this region constitute strong evidence in favor of the alternate hypothesis. Values outside this region provide strong evidence in favor of the null hypothesis. The values ± 1.96 are called *critical values*.

In our example, since $Z = -2.000$, the value of Z lies in the critical region and we reject the null hypothesis in favor of the alternate hypothesis.

We now consider another example.

Example 6.3.1. *A university medical study claims that the mean systolic blood pressure of male runners age 35–59 is 17 millimeters. We take a sample of 41 such runners and find their mean systolic blood pressure is 15.4 millimeters. Should we accept the claims of the university study?*

Solution: We again take $\alpha = .05$. Then

$$z = \frac{\bar{x} - \mu}{\sigma_{\bar{x}}}.$$

Further, $\sigma_{\bar{x}} = 17/\sqrt{41} = 2.65$.
Now,

$$Z = \frac{15.4 - 17}{2.65} = -0.6038.$$

Since $Z > -1.96$ and $Z < 1.96$, we accept the null hypothesis. $\square$

6.3.1 Errors

In our discussion of hypothesis testing we have seen that it is unlikely that the value of the test statistic falls in the critical region if the null hypothesis is true. That probability was only α, the significance level. But it is still possible for this to happen since α is not zero. If we reject the null hypothesis when it is actually true, we make what is typically called a *Type I error.* That is, a Type I error occurs when the test statistic falls in the critical region when the null hypothesis is actually true. Clearly,

$$P(\text{Type I error}) = \alpha.$$

Another type of error that can be made is to not reject the null hypothesis when it is false. This is called a *Type II error.* We always know α since we select the significance level. However, the probability of a Type II error is rarely known. This is because when the null hypothesis is false, we can no longer be sure the test statistic is normally distributed.

6.3.2   Summary of Hypothesis Testing

To summarize the procedure we have just outlined for hypothesis testing, we list the following steps in the process.

- Identify the null hypothesis and the alternate hypothesis. These are often conjectures or beliefs about certain values. Usually the null hypothesis involves a statement about equality.

- Choose the level of significance. Remember it is usually up to you to decide this parameter. Keep in mind, the more serious a Type I error would be, the smaller you should choose α.

- Determine the z value from the data. Before you do this, be sure that n is large enough.

- Determine the critical region. You need to know how to interpret the observed value.

- Make your decision on the null hypothesis.

6.3.3   One-Sided Tests

Sometimes our hypothesis test is really a one-sided test, that is, our critical region appears only on one side of the standard normal curve. As an example of such a test, consider the following:

Example 6.3.2. *A running coach claims that female runners tend to be taller than the average woman. The average woman is* 64 *inches tall. In a survey of* 45 *female runners we find they have a mean height of* $\bar{x} = 65.5$ *inches and that the sample standard deviation is* 3.5 *inches. Using this sample, do a* 5% *level of significance hypothesis test.*

Solution: We apply essentially the same five-step process we did for a two-sided test.

(1) Hypothesis: The claim is that $\mu > 64$. Hence, we use:

$$\mathcal{H}_0 : \mu = 64$$

$$\mathcal{H}_a : \mu > 64$$

for our null and alternate hypotheses.

(2) Level of significance: $\alpha = .05$.

(3) Test statistic and observed value:

$$Z = \frac{\bar{x} - \mu}{\sigma/\sqrt{n}} = \frac{65.5 - 64}{3.5/\sqrt{45}} = \frac{1.5}{.5217} = 2.875.$$

(4) Critical region: Since $\alpha = .05$, we have a critical point at $z = 1.65$. Thus, our critical region is $Z > 1.65$.

(5) Decision: Our observed value is much larger than 1.65 (not 1.96 since our region is on one side of the curve only). Hence, we reject the null hypothesis $\mathcal{H}_0$ in favor of the alternate hypothesis $\mathcal{H}_a$. This is the claim made by the running coach, that female runners are taller, on average, than the average over the entire population of women. $\qquad\square$

6.3.4 Small Sample Sizes

The fact that

$$Z = \frac{\bar{x} - \mu}{\frac{\sigma}{\sqrt{n}}}$$

is approximately standard normal for $n \geq 30$ allows us to use the strength of the normal distribution that we have established. But when the sample size is small, and here that means less than 30, we can no longer assume a standard normal distribution or take the approach we have used.

In 1908, while working for the Guiness Brewery in Dublin, Ireland, William Sealy Gosset became interested in statistical inference based upon small samples. This had directly to do with his work at the brewery. Gosset worked to develop a new test statistic he called t, which is defined as:

$$t = \frac{\bar{x} - \mu}{\frac{s}{\sqrt{n}}}$$

where s is the sample standard deviation, and n the sample size.

In an interesting historical twist, Gosset wrote about the t-statistic under the pen name of Student, since the brewery did not want its competition to know it had a statistician working for it! Because of this, the statistic has come to be called Student's t-statistic (see [39]).

We should note that there is not just one t-distribution, but infinitely many of them. Each one has a number associated with it called the *degree of freedom.* For our expression of t, the degree of freedom is $n - 1$.

A t-distribution resembles the standard normal distribution in shape. Its curve is symmetric with respect to a vertical line through 0 and it extends in both directions indefinitely. The expected value of t is 0. However, the t-distribution has a larger standard deviation, that is, it is more spread out than the standard normal distribution. Of course, the area under the curve is still one. As n grows large, the t curves approach the standard normal distribution. Appendix A.3 has a table of values for the probabilities of the t distributions we will need.

Example 6.3.3. *A sports energy drink claimed that every* 12 *ounce can of its drink contained* 500 *calories. To test the claim, a case of* 24 *cans were analyzed. It was found that* $\bar{x} = 507$ *calories and that for this sample,* $s = 21$ *calories. Test the claim at the* 2% *level of significance.*

Solution: The claim is that $\mu = 500$ calories. Hence we test

$$\mathcal{H}_0 : \mu = 500$$
$$\mathcal{H}_a : \mu \neq 500$$

with $\alpha = .02$.

Since the sample size is small, we must apply the t-test. Here we have
$$t = \frac{\bar{x} - \mu}{s/\sqrt{n}} = \frac{507 - 500}{21/\sqrt{24}} = 1.63.$$

Our degree of freedom is 23. Since our test uses "not equal" in $\mathcal{H}_a$, it is a 2-sided hypothesis test. From the table of Appendix A.3 we find $t_{\alpha/2} = t_{.01} = 2.5000$. Thus, the critical region consists of t values with $t \geq 2.5$ or $t \leq -2.5$. Then our value of t is not in the critical region, and so we do not reject $\mathcal{H}_0$.   □

Exercises:

6.3.1 Suppose an NBA coach claimed that on average, players shoot 45% from the field. Suppose a random sample of 36 NBA players showed a mean shooting percentage of .43 with a standard deviation of .04.

  1. Do a hypothesis test on the claim with a .05 significance level.

2. What is the probability of a Type I error?

3. What changes if the sample had been on 100 players instead of 36 players?

6.3.2 Suppose that same coach claimed that on average an NBA player shoots 65% from the free throw line. Suppose a sample of 36 random players showed a mean free throw shooting percentage of 68% with a standard deviation of 5%.

1. Do a hypothesis test on the claim with a .05 significance level.

2. What happens if the sample is on 100 players instead of 36 players?

3. What is the probability of a Type I error?

6.3.3 A team doctor felt his new treatment for sprained ankles was really helping heal players faster. Suppose a sprained ankle usually meant an average of 7 days without playing. The doctor observed the following days lost under his treatment: do a hypothesis test

$$
\begin{array}{cccccccccccc}
8 & 7 & 2 & 6 & 9 & 4 & 5 & 3 & 7 & 8 & 10 & 7 \\
7 & 6 & 4 & 10 & 3 & 6 & 8 & 2 & 5 & 4 & 4 & 4 \\
5 & 3 & 8 & 7 & 4 & 6 & 3 & 7 & 12 & 4 & 3 & 7
\end{array}
$$

with a .05 significance level on the doctor's claim.

6.3.4 An announcer claimed that NFL games, on average, seemed to take longer than 3 hours. Suppose a random sample of 49 games showed an average time of 3.2 hours with a standard deviation of .25 hours. Do a hypothesis test on the 3-hour claim with a significance level of .02.

6.3.5 A newspaper reports that the mean baseball salary is 3 million dollars per season. A random sample of 49 salaries showed a mean of 3.1 million dollars and a standard deviation of 3 million dollars.
(1) Do a .05 significance level hypothesis test on the claim.

(2) What would change if the random sample had been on 144 players?

6.3.6 Using Table 5.10 for the years 1992–2001, test, at the 98% confidence level, the hypothesis that Cal Ripken is a 20 home runs per season hitter.

6.3.7 Again using Table 5.10, for the period of 1982–1991, test, at the 95% level, the hypothesis that Cal Ripken is a 28 home runs per season hitter.

6.4   Are the Old Adages True?

Sports talk is filled with truisms, statements taken for fact by most, but hardly ever proven to be true. The home team has the advantage, always send up a right-handed hitter against a left-handed pitcher, always take the ball when you win the coin flip and many more such sayings. These sayings become ingrained in sports fans, most without real evidence of truth. There are many other things that are taken for granted that may or may not be true. Is it worthwhile bunting a runner to second base? In what situations is it advantageous to try to steal a base? When should you go for the 2-point conversion in football? Almost any strategic move in sports is open to debate (as can be seen nightly on any sports news show)!

In this section we look at a few of these adages and try to verify their validity statistically. If we can do that, then they become facts rather than adages.

6.4.1   Home Field Advantage

We begin with the easiest adage to check, home field advantage. You hear this in every sport. Playing at home is better than playing on the road. Reasons are sometimes given like "the players are more comfortable in their surroundings." Such reasons are impossible for us to test.

We will test this adage with the data in Table 6.2 and Table 6.3. These tables show the home and away records for all American League teams in 2008 and 2007.

TABLE 6.2:   2008 Home and Away
Records for AL

Team	Home Record	Away Record
Tampa Bay	57-24	40-41
Boston	56-25	39-42
New York	48-33	41-40
Toronto	47-34	39-42
Baltimore	37-43	31-50
Chicago	54-28	35-46
Minnesota	53-28	35-47
Cleveland	45-36	36-45
Kansas City	38-43	37-44
Detroit	40-41	34-47
Los Angeles	50-31	50-31
Texas	40-41	39-42
Oakland	43-38	32-48
Seattle	35-46	26-55

When we examine this data we see some teams have winning home
and losing road records, but others have losing records both at home
and away. Thus, it is insufficient to just observe the home and away
records. What we will consider is the home winning percentage minus
the road winning percentage. If this value is positive, then the team
has performed better at home than away. For the 2008 season, these
differences are shown in Table 6.4.

After examining the differences, we see that none are negative (two
were 0), so no team performed worse at home than away, while almost all teams performed better. Our conclusion then is that the data
supports the claim that home field advantage is real.

Now we will test a hypothesis on home field advantage. Let $\mathcal{H}_0$ be
that home teams win in the American League at a .54 rate and $\mathcal{H}_a$ be

TABLE 6.3: 2007 Home and Away
Records for AL

Team	Home Record	Away Record
Boston	51-30	45-36
New York	52-29	42-39
Toronto	49-32	34-47
Baltimore	35-46	34-47
Tampa Bay	37-44	29-52
Cleveland	52-29	44-37
Detroit	45-36	43-38
Minnesota	41-40	38-43
Chicago	38-43	34-47
Kansas City	35-46	34-47
Los Angeles	54-27	40-41
Seattle	49-32	39-42
Oakland	40-41	36-45
Texas	47-34	28-53

that American League home teams win at a higher rate.

We wish to test this hypothesis. We will use our two years of data for this test. Note that the mean home winning percentage for these American League teams over the 2007–2008 seasons is $\bar{x} = .5590$ with a standard deviation of .0862. Finding z we see that

$$Z = \frac{.5590 - .54}{.0862/\sqrt{28}} = .5549.$$

This Z value is well below 1.96 and so we are 95% sure that $\mathcal{H}_0$ holds and we accept the null hypothesis. That is, we accept the hypothesis that American League teams win at home with a mean rate of .54.

When you check recent results, all major league baseball teams on average have won approximately 52% of their home games, football teams have won 58% of their home games, while basketball teams have won 66% of their home games (see [2]). So these other sports have an

TABLE 6.4:   Percentage Differences for 2008

Team	Difference Computation	Team	Difference Computation
Tampa Bay	.7037 - .4938 = .2099	Boston	.6913 - .4814 = .2096
New York	.5925 - .5062 = .0864	Toronto	.5802 - .4814 = .0988
Baltimore	.4568 - .3827 = .0741		
Chicago	.6667 - .4321 = .2346	Minnesota	.6543 - .4321 = .2222
Cleveland	.5556 - .4445 = .1111	Kansas City	.4691 - .4691 = .0000
Detroit	.4938 - .4198 = .0740		
Los Angeles	.6173 - .6173 = .0000	Texas	.4938 - .4815 = .01213
Oakland	.5309 - .3951 = .1358	Seattle	.4321 - .3210 = .1111

even stronger home field advantage. Overall, the evidence supports the adage that home field advantage is real.

6.4.2   Lefty Versus Righty

Next we wish to look at another old baseball adage that hitters hit better against a pitcher of the "opposite hand," that is, right-handed hitters will hit better against left-handed pitchers than against right-handed pitchers and left-handed hitters will hit better against right-handed pitchers than left-handed pitchers.

Our approach will be similar to the one we took for home field advantage, that is, we will find the batting averages of a number of players against both left- and right-handed pitching. We will then look at the difference between the player's average against the opposite hand pitching minus their average against the same hand pitching. Table 6.5 contains the data for 15 players selected from National League East teams. Three players from each team were selected. No switch hitters were selected, since they never face "same handed" pitching. Also, only players with a significant number of at bats were considered. Data was collected on each player for both 2007 and 2008.

When we ask a question like "Is there a home field advantage?" or

"Do right-handed hitters perform better against left-handed pitchers?"
we are asking if a special situation has an effect on outcomes. There
are essentially three possible answers to such a question:

(1) There is no effect from the special situation.

To draw this conclusion we should see essentially no difference in
outcomes when the special situation holds and when it does not hold.

(2) There is a general effect from the situation.

To draw this conclusion we should see evidence that outcomes are
different during the special situation.

(3) There is a situational effect, but it depends on another feature such
as a player's ability.

To draw such a conclusion we should see evidence that only a certain
few players have enhanced results during the special situation.

For the home versus away situation, we saw that in our sample data
all teams did at least as well at home as away and the vast majority
actually performed better at home than away. Thus, we concluded that
home field advantage is real.

Now for the adage of opposite armed pitchers, we look at the *observed
situational effect*; that is, we consider the difference between the hitters'
batting average against pitchers of opposite arm minus their batting
average against pitchers of the same arm. Hence, we find

$$diff = average_{opp\ arm} - average_{same\ arm}.$$

These differences are shown in Table 6.5. Considering each of the 15
players, over the two years, we have 30 differences. Of these, 7 were
negative (5 in 2008 and 2 in 2007). Further, no player had a negative
difference in both years. Thus, the evidence is not as overpowering as
for home field advantage.

When we average the differences in 2008 we see that the mean difference was .0167 while for 2007 the mean difference was .050. In [2],
it was claimed that players' batting averages are around .015 higher
against pitchers of the opposite arm. Our results were approximately
that for 2008 and higher yet for 2007. Of course, we were working with
a fairly small sample of size 30, not all players over both years.

But what we can conclude is that our sample supports answer (2),
that there is generally a positive effect; that is, hitters tend to perform
better against pitchers of the opposite arm.

TABLE 6.5:   2007–2008 Differences in Averages

Name	Bats	2008 Lefty	Righty	Diff.	2007 Lefty	Righty	Diff.
David Wright	R	.382	.275	.107	.361	.311	.050
Carlos Delgado	L	.267	.273	.006	.267	.254	-.013
Damon Easley	R	.287	.260	.027	.371	.202	.169
Lastings Milledge	R	.258	.272	-.014	.317	.250	.067
Ryan Zimmerman	R	.333	.259	.074	.374	.235	.139
Ronnie Belliard	R	.307	.279	.038	.329	.275	.054
Brian McCann	L	.299	.301	.002	.264	.273	.009
Jeff Francoeur	R	.210	.251	-.041	.317	.281	.036
Yunel Escobar	R	.282	.299	-.017	.355	.303	.052
Chase Utley	L	.277	.301	.024	.318	.340	.022
Pat Burrell	R	.279	.238	.041	.255	.257	-.002
Jayson Werth	R	.303	.255	.048	.375	.257	.118
Dan Uggla	R	.191	.283	-.092	.245	.245	.000
Jeremy Hermida	L	.240	.252	.012	.282	.297	.015
Hanley Ramirez	R	.258	.313	-.055	.399	.312	.047

Next, we will use our data to test the hypothesis that hitters on average hit .015 better against pitchers of the opposite arm. Thus,

$\mathcal{H}_0$ is on average, hitters hit .015 points higher against pitchers of the opposite arm.

$\mathcal{H}_a$ is that hitters hit higher than .015 better against pitchers of the opposite arm.

Using the data from Table 6.4 we find a mean difference in batting average of .0307 with a standard deviation of .0051.

Computing Z for a .05 significance level hypothesis test we obtain

$$Z = 1.5678.$$

But $Z = 1.5678 < 1.65$ and so Z does not lie in the critical region for the one-sided test of the null hypothesis. Thus, we accept the null hypothesis in this case. That is, our data supports the claim that hitters, on average, hit about .015 points higher against pitchers of the opposite arm.

Exercises:

6.4.1 Suppose NFL football teams average 37 touchdowns per 16-game season. Also suppose 96% of extra point kicks are successful while only 50.9% of 2-point conversions are successful. Compare the expected number of points per game of a team that always kicked an extra point versus a team that always tried for a 2-point conversion. What changes if the 2-point conversion success rate is only 45%?

6.4.2 Suppose an NBA team takes, on average, 100 shots per game. Also suppose the team makes 45% of the 2-point shots taken and 30% of the 3-point shots taken. The team normally takes eighty 2-point shots and twenty 3-point shots. Would the team be better off taking fewer 3-point shots? Would the team be better off taking only 3-point shots? Would the team be better off taking only 2-point shots?

6.4.3 The chart below is a part of the Expected Runs Table for the $1977 - 1992$ MLB seasons (see [20]). It shows certain typical situations and the expected number of runs that a team would score in that situation. Does the data support the claim that with no outs and a runner on first base, sacrifice bunting a runner to second base is a good baseball strategy?

Runners on	AL	0	1	2	NL	0	1	2
_ _ _		.498	.266	.099		.455	.239	.090
x _ _		.877	.522	.224		.820	.490	.210
_ x _		1.147	.693	.330		1.054	.650	.314

6.4.4 The chart below is a part of the One Run Probability Table for the 1977–1992 MLB seasons (see [20]). It measures the probability of scoring one run in a given situation. Does the chart support the claim that with no outs and a runner on first, sacrifice bunting the runner to second base is a good play?

Runners on	AL	0	1	2	NL	0	1	2
_ _ _		.276	.161	.067		.261	.148	.061
x _ _		.432	.277	.129		.424	.268	.124
_ x _		.634	.414	.226		.609	.400	.216

6.5 How Good Are Certain Measurements?

We have discussed a number of different sports statistics and drawn a number of different conclusions. We have stated that fans generally do not think the best batting average makes someone the best hitter. Also, the most home runs does not make someone the best hitter. But are these statistics useful measures at all? Are these measures really important to winning?

Note: HR = home runs, BA = batting average, OBP = on base percentage, OPS = on base plus slugging percentage.

It can be argued that the most important offensive statistic for any baseball team is runs scored. You need to score more than your opponent to win, so runs are clearly critical. Also, runs are produced in a variety of ways, not just from one individual hit. Thus, scoring runs seems more a function of the team batting performance, not just that of an individual.

In this section we want to correlate several common baseball statistics with runs scored. If we find there is a reasonable correlation between

TABLE 6.6: 2008 Team Offensive Statistics

Team	Runs	HR	BA	OBP	Slug	OPS
Texas	901	194	.283	.354	.462	.816
Boston	845	173	.280	.358	.447	.805
Minn	829	111	.279	.340	.408	.748
Detroit	821	200	.271	.340	.444	.784
Chi Sox	811	235	.263	.332	.448	.780
Clev	805	171	.262	.339	.424	.763
NYY	789	180	.271	.342	.427	.769
Balt	782	172	.267	.333	.429	.762
Tam Bay	774	180	.260	.340	.422	.762
LA A	765	159	.268	.330	.413	.743
Toronto	714	126	.264	.331	.399	.731
KC	691	120	.269	.320	.397	.717
Sea	671	124	.265	.318	.389	.707
Oak	646	125	.242	.318	.369	.686
ChiCubs	855	184	.278	.354	.443	.797
NY Mets	799	172	.266	.340	.420	.761
Phil	799	214	.255	.332	.438	.770
St. L	779	174	.281	.350	.433	.783
Flor	770	208	.254	.326	.433	.759
Atl	753	130	.270	.345	.408	.753
Mil	750	198	.253	.325	.431	.757
Col	747	160	.263	.336	.415	.751
Pitt	735	153	.258	.320	.403	.723
Ariz	720	159	.251	.327	.415	.742
Houston	712	167	.263	.323	.415	.737
Cincin	704	187	.247	.321	.408	.729
LA D	700	137	.264	.333	.399	.732
Wash	641	117	.251	.323	.373	.696
SF	640	94	.262	.321	.382	.703
SD	637	154	.250	.317	.390	.707

these statistics and runs scored, then there is a stronger justification for recording these statistics. To help us, we will use the 2008 team offensive statistics shown in Table 6.6. For each team in major league baseball we have recorded the runs scored, home runs hit, team batting average, and team on base percentage.

We begin by plotting home runs versus runs scored for the American League teams in 2008, as well as the least squares approximating line for these points. This plot is shown in Figure 6.2.

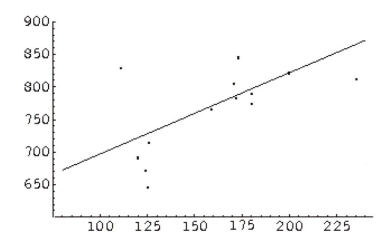

FIGURE 6.2: Home runs versus runs scored, AL 2008.

Calculating the least squares approximating line we find it is

$$y = 1.2547x + 572.451.$$

When we compute the correlation coefficient for the American League data of home runs versus runs scored, we obtain a correlation coefficient of

correlation coefficient = .6710.

This shows a mild correlation between home runs hit and runs scored. Recall, correlation coefficients close to 1 indicate a strong correlation while those close to 0 indicate almost no correlation. We are slightly above the middle of that range, hence we have only a mild correlation. This tends to support the theory that there is usually more to scoring a run than just hitting a home run. But this was a small data set.

Now we include the data from the National League teams. The points for all Major League teams in 2008 are shown in Figure 6.3 as well as the least squares approximating line. Note that this approximating line is

$$y = 1.2489x + 549.757.$$

Thus, the addition of the extra data points does very little to change the approximating line. The slope is essentially the same. When we compute the correlation coefficient for the expanded data set for home runs versus runs scored for both leagues we obtain

$$\text{correlation coefficient } = .6369.$$

Thus, the larger data set remains only mildly correlated and in fact the coefficient has decreased slightly. Again this supports the claim that home run hitting and runs scored are not strongly correlated. We certainly have not proven that fact, as our data covers only one year. But we have provided some reasonable evidence of the claim.

6.5.1   Batting Average and Runs Scored

Next we turn to another potential correlation question, namely team batting average versus runs scored. We approach this question in a manner analogous to that for home runs versus runs scored. We again use the 2008 data for our test.

For the team batting averages in 2008, of the 30 major league baseball teams versus runs scored, and the least squares approximating line, see Figure 6.4.

This time we see a more defined shape to the point plot. We also find the least squares approximating line to be

$$y = 4707x - 491.803.$$

The large slope is caused by the difference in scale between batting averages and runs. The real test is in the correlation coefficient. This time the correlation coefficient for team batting average versus runs scored is

$$\text{correlation coefficient } = .9704.$$

Thus, batting average is very strongly correlated with runs scored (at least in this data set). Again this is not a proof of the fact, as the data is only for 2008. However, it is very good evidence that batting average is strongly correlated to runs scored.

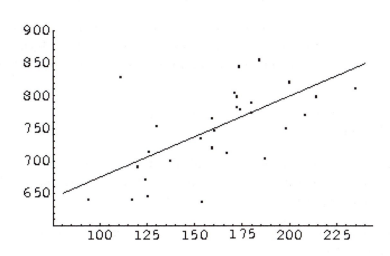

FIGURE 6.3:   Home runs versus runs scored, all of MLB 2008.

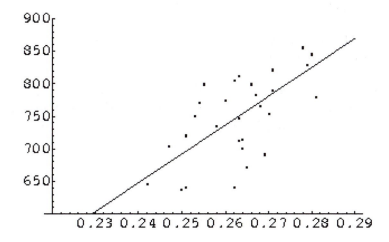

FIGURE 6.4:   Team batting averages versus runs scored, MLB 2008.

Exercises:

6.5.1 Using the 2008 data supplied in this section, plot On Base Percentage (OBP) versus runs scored for the American League.

6.5.2 Using the data of the previous exercise, compute the correlation coefficient for OBP versus runs scored from this data. If the correlation coefficient justifies it, find the least squares approximating line.

6.5.3 Plot slugging percentage versus runs scored for the National League.

6.5.4 What is the correlation coefficient for the data of the previous problem? If the correlation coefficient justifies it, find the least squares approximating line.

6.5.5 Plot OPS versus runs scored for the American League using the 2008 data.

6.5.6 What is the correlation coefficient for the data of the previous exercise?

6.5.7 Plot home runs versus batting average for the National League using the 2008 data.

6.5.8 What is the correlation coefficient for the data of the previous problem? If the correlation coefficient justifies it, find the least squares approximating line.

6.6 Arguing over Outstanding Performances

Sports fans are always arguing over the best performance of some player or players. If we are faced with a given statistic for a particular player, it is sometimes difficult to judge the merits of this statistic because we do not know the context in which it was achieved. For example, Carl Yastrzemski hit .301 for the Red Sox in 1968. Now, .301 is a good average, but it is not necessarily one that would cause a great stir. However, Yastrzemski won the American League batting title that year with his .301 average. The mean for the American League was only .230 in 1968.

In 1980 Alan Trammel hit .300 for the Tigers, very comparable to Yastrzemski's .301. But in 1980, Mickey Rivers hit .333 and in fact, 18 players hit better than Trammel, with George Brett hitting .390, a full 30% better than Trammel's .300 average. Thus, the context in which a performance is realized is meaningful in measuring the achievement.

We have the tools already to measure a performance relative to the particular context in which it was achieved. A natural approach is to compute the standard Z score for the performance. As you will recall, the Z score gives a measure of how many standard deviations a particular statistic is from the mean. Note the standard deviation in both 1977 and 1980 was approximately .027.

For example, in 1977 Rod Carew hit .388 to lead the American League. Carew's Z score for 1977 is

$$Z_{Carew} = \frac{.388 - .26622}{.027} = 4.51.$$

To be more than 4 standard deviations above the mean is a truly outstanding performance for Carew.

For 1980, George Brett's Z score is

$$Z_{Brett} = \frac{.390 - .26907}{.027} = 4.47.$$

Again an impressive performance by Brett.

Mickey Rivers Z score for 1980 is 2.37 and Alan Trammel's Z score is 1.14. These numbers help put a perspective on each of their relative positions in the league that year.

In 1994, Tony Gwynn hit .394, coming as close to being a .400 hitter as anyone has in the years since Ted Williams .406 in 1941. Gwynn's Z score was 4.70, thus even farther separated from the mean as was Brett or Carew.

In trying to measure these outstanding performances we run into a problem in trying to compare them. These Z scores are relative to their own league in a particular year. Comparing two different years or leagues this way does not make complete sense. So something other than the Z score is needed in the argument as to which of Carew, Brett or Gwynn had the best year. Maybe Yastrzemski's 1968 performance is actually more impressive (his Z score is about 2.6).

Since sports fans are always arguing between leagues and across years, some other methods of comparison have developed. An early suggestion by David Shoebothom and Merritt Clifton (see [35]) was called the *relative batting average* (RBA). The computation for the RBA is

$$\text{RBA} = \frac{\text{player average}}{\text{league average}} \times 100.$$

Hence, *RBA* is just that, a measure of batting relative to the league.

For George Brett in 1980 we obtain

$$\text{Brett}_{RBA} = \frac{.390}{.269} \times 100 = 145.00.$$

How do we interpret this number? An RBA of 100 means the player's batting average was exactly that of the league. Brett's RBA should be interpreted as 45% above the league average. But again, RBA is still relative to that particular league, in that particular year.

In [35], Michael J. Schell suggests a variety of adjustments to help compare such statistics. (More than we have time for, as his entire book is about these adjustments.) One such adjustment that is easy to apply is the Mean-Adjusted Batting Average (MABA).

To compute MABA you apply the following formula:

$$\text{MABA} = \frac{\text{player average}}{\text{league average}} \times \text{league mean.}$$

For the period 1969–1992, the league mean was .255 (see [35]). Thus, for George Brett we obtain an MABA of

$$\text{Brett}_{MABA} = \frac{.390}{.269} \times .255 = .370$$

while for Rod Carew,

$$\text{Carew}_{MABA} = \frac{.388}{.266} \times .255 = .372$$

and for Tony Gwynn,

$$\text{Gwynn}_{MABA} = \frac{.394}{.267} \times .255 = .376.$$

Now, we have a more meaningful argument as to which player had a better year as each has been adjusted against the mean of his league and the mean of both leagues over a longer period.

As suggested by Schell [35], a combination of many other adjustments are possible in this argument. Schell also used a standard deviation adjustment as follows:

$$\text{SD adjustment} = MN + \frac{.029}{SD} \times (MABA - MN).$$

Here MN and SD are the mean and standard deviation of the MABA values for all players with sufficient at bats and .029 is the standardized standard deviation. Clearly, this requires a great deal more overall computation of values to apply.

Exercises:

6.6.1 Find the RBA for Carl Yastrzemski in 1968.

6.6.2 Find the RBA for Alan Trammel in 1980.

6.6.3 Find the RBA for Mickey Rivers in 1980.

6.6.4 Find the $MABA$ for Carl Yastrzemski in 1968. You may use .255 as the league mean.

6.6.5 Find the $MABA$ for Alan Trammel in 1980.

6.6.6 Find the $MABA$ for Mickey Rivers in 1980.

6.7   A Last Look at Comparisons

We end this chapter with one more look at comparisons of means, this time when both samples are small. There are many sports situations where such comparisons can be applied. The technique is a specialized case of the t-statistic.

6.7.1   Small Sample Comparisons

This section is based upon yet another comparison of Albert Pujols and Ichiro Suzuki. We considered these two players in some detail earlier. In this section we compare them on a hits-per-season basis.

Each player has played 8 seasons (use our earlier data). Their hit totals for each season are shown in the stemplot of Table 6.7. We base our test on the hypothesis that the mean number of hits per season for Pujols and that for Suzuki are equal. That is,

$$\mathcal{H}_0 \; : \; \mu_P = \mu_S$$

$$\mathcal{H}_a \; : \; \mu_P \neq \mu_S.$$

TABLE 6.7:   Stemplot of Hits per Season, Pujols Versus Suzuki

Pujols		Suzuki
7	17	
755	18	
654	19	
	20	68
2	21	23
	22	4
	23	8
	24	2
	25	
	26	2

From the stemplot data we determine that $\bar{x}_P = 191.375$ and $\bar{x}_S = 225.625$. Further, the sample standard deviations are $s_P = 10.5144$ and $s_S = 19.9566$. We also have that $n_1 = n_2 = 8$.

In comparing the means of two small samples, one uses the t-statistic

$$t = \frac{(\bar{x}_1 - \bar{x}_2) - (\mu_1 - \mu_2)}{\sqrt{s_1^2/n_1 + s_2^2/n_2}}.$$

where $\bar{x}_i$ are the sample means, μ_i the population means, and s_i the sample standard deviations, $i = 1, 2$. This statistic may be treated as having a Student's t distribution with degrees of freedom equal to the smaller of $n_1 - 1$ and $n_2 - 1$. The value for $\mu_1 - \mu_2$ is obtained from the null hypothesis and a $1 - \alpha$ confidence interval for $\mu_1 - \mu_2$ is

$$(\bar{x}_1 - \bar{x}_2) \pm t_{\alpha/2}\sqrt{s_1^2/n_1 + s_2^2/n_2}.$$

Now suppose our hypothesis is that Pujols and Suzuki average the same number of hits per season, that is, $\mu_P = \mu_S$. Our hypothesis test again is

$$\mathcal{H}_0 : \mu_P = \mu_S$$

$$\mathcal{H}_a : \mu_P \neq \mu_S.$$

From our two data sets we have that $\bar{x}_P = 191.375$ and $\bar{x}_S = 225.625$. The degrees of freedom here are 7. Thus, from Table 9.2 we see that $t_{.025} = 2.365$. Hence,

$$(\bar{x}_P - \bar{x}_S) - 2.365\sqrt{s_1^2/n_1 + s_2^2/n_2} < \mu_P - \mu_s$$

and

$$\mu_P - \mu_S < (\bar{x}_P - \bar{x}_S) + 2.365\sqrt{s_1^2/n_1 + s_2^2/n_2}$$

Upon substituting we obtain

$$-53.1111 < \mu_P - \mu_S < -15.3889$$

that is, our interval is $[-53.1111, -15.3889]$. Thus, the confidence interval for the value we seek shows this difference to be negative (with very high probability).

The t-statistic we seek is actually

$$t = \frac{(\bar{x}_P - \bar{x}_S) - (\mu_P - \mu_S)}{\sqrt{s_1^2/n_1 + s_2^2/n_2}}.$$

Now $\mu_P - \mu_S$ is a part of this and we do not know this value precisely. What we shall do is assume the null hypothesis is true until we show it is not. That is, we substitute the assumed value of $\mu_P - \mu_S$ into the formula. In our present case this value is zero, as we assumed $\mu_P = \mu_S$.

Thus, we obtain

$$t = \frac{(\bar{x}_P - \bar{x}_S)}{\sqrt{s_1^2/n_1 + s_2^2/n_2}}$$

or

$$t = -4.29461.$$

But this t value is in the critical region $t \le -2.365$ or $t \ge 2.365$. Hence, we must reject the null hypothesis.

It is not a major surprise that we rejected the null hypothesis in this case. Suzuki had almost 300 more hits in his 8 seasons, so his mean number of hits per season is significantly higher than that of Pujols. Suppose we try a more reasonable null hypothesis. This time we will test

$$\mathcal{H}_0 \ : \ \mu_P - \mu_S = -25$$

$$\mathcal{H}_a \ : \ \mu_P - \mu_S \ne -25.$$

Now when we compute the t-statistic we obtain

$$t = -1.15986$$

which is not in the critical region for a 95% test. Hence, in this case we accept the new null hypothesis that Suzuki gets 25 hits more per season.   □

Exercises:

6.7.1 Using Table 5.12, find a 95% confidence interval on the average number of points scored per season of Jerry West and Oscar Robertson being equal using the data over the first 8 years of the table.

6.7.2 Do the same test as in the previous problem, but over the last 8 years of the table.

6.7.3 Test the hypothesis that Robertson and West averaged the same number of points per season over the first 8 years of data from Table 5.12 at the 98% level.

6.7.4 Test the same hypothesis as in the previous problem, but using the data of the last 8 years of the table.

6.7.5 Find two NFL quarterbacks with at least 5 years experience. Find a 95% confidence interval for the difference in their average number of TD passes based on 5 years worth of data for each player.

6.7.6 Using the data of the previous problem, test at the 95% confidence level, the hypothesis that the two means are equal.

6.7.7 Select any two NFL running backs with at least 5 years experience. Find a 95% confidence interval on the value of the difference in their mean yearly yardage gained by rushing.

6.7.8 Run a 95% hypothesis test on the means of the running backs of the previous problem being equal.

6.7.9 Suppose one survey of football injuries in games played on artificial turf showed that over 20 games the average number of injuries per game was 15.2 with a standard deviation of 2.2. Another survey of injuries in games played on grass showed that over 20 games the average number of injuries per game was 13.8 with a standard deviation of 1.8. Does this data support the claim that there are more injuries on artificial turf than on grass?

Chapter 7

Games and Puzzles

7.1 Introduction

In this chapter we will look at a number of games and puzzles that have interesting mathematical ties. Each is a real game or puzzle and most have been or still are being marketed. Some of these games come with a long history (magic squares) or a tall tale (Tower of Hanoi). While others have supposed mystic properties (magic squares). Some of the games are electronic (Lights Out), others are plastic (Instant Insanity), some are wood (peg games), others require only paper and pencil (Sudoku). Some have become uncommonly popular (Sudoku).

No matter which game or puzzle we consider, our approach will be the same. Find the underlying mathematics and use it to our advantage in the game! Hopefully as a means of solving the puzzle or winning the game.

7.2 Number Arrays

There are several puzzles that require you to fill in the entries of a $n \times n$ (square) array with integers satisfying particular properties. In this section we will look at two of these puzzles.

7.2.1 Magic Squares

The first puzzle we consider is called a *magic square*. A magic square is an $n \times n$ array of positive integers whose entries are the integers $1, 2, \ldots, n^2$ arranged so that the sum of any row, column or main diagonal is the same value. This value is called the *magic constant*.

The first known example of a magic square (see Table 7.1) is taken from the *Loh-Shu* scroll (or scroll of the river Lo) in China. Scholars date it to the mythical founder of Chinese civilization, Fuh-Shi, around 2858–2736 BC. But magic squares have been found in drawings from many other cultures including Egypt and India (see [21]). An Arab mathematician, Ahmad al-Buni, worked on magic squares around 1200 AD and he attributed mystical properties to them. However, the details of these properties are not known today. In addition, there are references to the use of magic squares in astrological calculations (see [21]).

TABLE 7.1: Loh Shu Magic Square

4	9	2
3	5	7
8	1	6

Before we consider finding the magic constant, the following result on binomial coefficients will be helpful.

Theorem 7.2.1. *The sum of the first m positive integers is* $\binom{m+1}{2}$, *that is,*

$$\sum_{i=1}^{m} i = \binom{m+1}{2}.$$ (7.1)

Proof: In order to see this result, first suppose m is even, say $m = 2k$. Then,

$$\sum_{i=1}^{m} i = 1 + 2 + \ldots + k + (k+1) + \ldots + 2k - 1 + 2k$$
$$= (1 + 2k) + (2 + (2k-1)) + \ldots + (k + (k+1))$$
$$= (2k+1)k$$
$$= (2k+1)(2k)/2$$
$$= (m+1)(m)/2$$
$$= \binom{m+1}{2}.$$

If m is odd, then using what we have already shown, just add the first $m-1$ terms first to obtain $\binom{m}{2}$, then add in m to this. Again the sum will be $\binom{m+1}{2}$. ☐

Example 7.2.1. *Given an $n \times n$ magic square, what is the magic constant?*

Solution: In the case of an $n \times n$ magic square, if we add all the entries and average them over the n rows, we will obtain the magic constant. That is, using Equation 7.1

$$\frac{1}{n}\sum_{i=1}^{n^2} i = \frac{\binom{n^2+1}{2}}{n} = \frac{n(n^2+1)}{2}.$$

Thus, the magic constant for arbitrary n is

$$M_c = \frac{n(n^2+1)}{2}.$$

☐

Note that if $n = 1$, then the magic constant is $M_c = 1$.

If $n = 2$, then the magic ocnstant is $M_c = 5$. Thus, the rows of the array must be $1, 4$ and $2, 3$ in some order. But with these rows, there is no way for the columns to sum to 5. Hence, there can be no 2×2 magic square.

For $n = 3$, $M_c = \frac{3(10)}{2} = 15$. With this in mind we find that one such square is the one shown in Table 7.2

TABLE 7.2: A Magic Square

8	1	6
3	5	7
4	9	2

We can easily check that the rows, columns and main diagonals of this square all sum to 15.

If we form a new square array by taking each term of a magic square and subtracting that term from $n^2 + 1$, we obtain a new magic square called the *complementary square*. In the case that $n = 3$ we see that $n^2 + 1 = 10$. Thus, the complementary square for the square of Table 7.2 is the one shown in Table 7.3. You should compare this square with the original. Can you describe what has happened in transforming the magic square to its complement?

TABLE 7.3:　The Complementary Square

2	9	4
7	5	3
6	1	8

Example 7.2.2. *Justify why the subtraction of entries from $n^2 + 1$ produces another magic square.*

Solution: Consider any row (column or main diagonal) with entries

$$x_1, x_2, \ldots, x_n.$$

Then the sum of the entries in the complementary square is

$$\sum_{i=1}^{n}((n^2+1)-x_i) = n(n^2+1)-\sum_{i=1}^{n}x_i = n(n^2+1)-\frac{n(n^2+1)}{2} = \frac{n(n^2+1)}{2}.$$

Hence, we obtain the magic constant again, and so, since the same argument holds for any column or main diagonal, the complementary square is also a magic square.　□

It is an unsolved problem to determine the number of distinct magic $n \times n$ squares (excluding those obtained by rotation or reflection). However, for small values of n these numbers are known (see Table 7.4).

Problem 7.2.1. *Construct a 4×4 magic square.*

Question 7.2.1. *How can we attempt to find magic squares?*

TABLE 7.4: The Number of Small Magic Squares

n	number of magic squares
1	1
2	0
3	1
4	880
5	275305224

TABLE 7.5: An Unknown 3×3 Magic Square

a	b	c
d	e	f
g	h	i

There is a solution method using linear algebra, but we will try to accomplish the same result with a little less mathematical firepower. Suppose we want a 3×3 magic square, say the one in Table 7.5.

We know $M_c = 15$. If we subtract $\frac{n^2+1}{2} = 5$ from each entry, we obtain a 3×3 array with entries

$$-4, -3, \ldots, -1, 0, 1, \ldots, 3, 4.$$

Further, the row, column and main diagonal sums are now all zero. We can use this information to build a system of equations (see Table 7.6).

TABLE 7.6: System of 8 Equations in 9 Unknowns

rows	columns	diagonals
$a + b + c = 0$	$a + d + g = 0$	$a + e + i = 0$
$d + e + f = 0$	$b + e + h = 0$	$c + e + g = 0$
$g + h + i = 0$	$c + f + i = 0$	

Thus, we have a system of 8 equations in 9 unknowns. If we had at least as many equations as unknowns, we know we could solve the system. We could add a suitable multiple of the first equation to the second equation to get a new equation not involving one of the variables, say x. Similarly, we could add multiples of the first equation to the third, fourth, ... , mth equations getting $m - 1$ equations, none of which involve this variable x. Repeating this technique, eliminating another variable, say y, and so on, we get down to just one unknown. We can solve for this and back substitute to find the rest.

But all is not lost here. It turns out that the equations are not all linearly independent. To see this we will need to solve for some of the variables in terms of other variables.

Solving the row equations for a, d and g, respectively, and plugging these values into the first column equation we obtain column equations which sum to zero.

$$
\begin{array}{rrrrrrl}
-b & -c & -e & -f & -h & -i & = 0 \\
+b & & +e & & +h & & = 0 \\
& +c & & +f & & +i & = 0
\end{array}
$$

Therefore, $a+d+g = 0$ is redundant to the other 2 column equations. Now, after substituting for a and g in the 2 diagonal equations, we are left with 4 equations (the last 2 column equations and 2 diagonal equations) in 6 unknowns.

$$
\begin{array}{rllll}
(1) & b & +e & +h & = & 0 \\
(2) & c & +f & +i & = & 0 \\
(3) & & +e & +i & = & b & +c \\
(4) & c & +e & & = & h & +i
\end{array}
$$

Given two of the unknowns, say h and i, we can solve for the rest. Adding equations (3) and (4) we obtain

$$2e = b + h.$$

This implies

$$3e = b + h + e = 0.$$

Which in turn implies

$$e = 0.$$

But now we can say that

$$b = -h$$
$$c = h + i$$
$$f = -c - i = -(h + i) - i = -h - 2i.$$

Further, we know that the sum of the squares of all the variables is 60. That is,

$$a^2 + b^2 + \ldots + i^2 = (-4)^2 + (-3)^2 + \ldots + (3)^2 + (4)^2 = 60.$$

Substituting for b, c and f in the left-hand part of this equation produces

$$2i^2 + 2hi + h^2 = 10.$$

This is the equation for an *ellipse*. It is straightforward to verify that the eight integer points

$$(-3, 2), \quad (-3, 4), \quad (-1, 4), \quad (1, 2)$$
$$(3, -2), \quad (3, -4), \quad (1, -4), \quad (-1, -2)$$

all satisfy this equation (recall our values run from -4 to 4, so these turn out to be all the integer pairs in this range that satisfy the equation). From this, for example, we can assign the point $i = -1$ and $h = 4$. Now plugging these values into the variables produces the magic square we initially gave as an example. Using the other seven points for i and h we obtain the eight rotations or reflections of a 3×3 magic square, giving all possible ways of producing such a square. Thus, we can consider this a complete solution for the 3×3 magic squares.

7.2.2 Variations on Magic Squares

There is a simple relaxation of the magic square properties that still allows some interesting results. We will call a square *weakly magic* provided the row and column sums are all equal to the magic constant. Thus, we no longer require the diagonals to also sum to the magic constant.

As a convenience in our construction, we consider the following simple variation. We will use the entries 0 to $n^2 - 1$ for our weakly magic squares. This simple variation allows us to take advantage of some arithmetic rules.

Definition 7.2.1. *Suppose that a, b and n are integers with $n > 0$. We say that a and b are congruent modulo n if and only if n divides $(a - b)$. We denote this relationship as*

$$a \equiv b \bmod n$$

and read this as a is congruent to b modulo n.

As a simple example, note that the integers 35 and 15 are congruent modulo 10, since

$$35 - 15 = 20 = 2 \times 10.$$

Similarly, each of 35 and 15 is congruent to 5 modulo 10, since

$$35 - 5 = 30 = 3 \times 10$$

and

$$15 - 5 = 10.$$

Thus, $35 \equiv 5 \bmod 10$ and $15 \equiv 5 \bmod 10$. □

Clearly, any positive integer is congruent to its remainder when divided by n, that is, saying

$$x \equiv r \bmod n$$

is equivalent to saying

$$x = cn + r$$

where $r < n$ and c is a nonnegative integer.

We will find this arithmetic useful on our modified squares, since any integer is congruent to one of $0, 1, \ldots, n - 1$ modulo n and these are our entries in the modified weak magic squares.

Weakly magic squares have ties to other mathematical structures. For example, suppose we have the weakly magic square of Table 7.7.

A *Latin Square* is an $n \times n$ array with entries from $\{1, 2, \ldots, n\}$ such that each integer appears exactly once in each row and each column (see Table 7.8). Further, two Latin Squares are said to be *orthogonal*

TABLE 7.7: A Weak Magic Square
with Modified Entries

0	5	10	15
7	2	13	8
9	12	3	6
14	11	4	1

provided the pairs obtained from entries with the same coordinates are all distinct and hence all the possible pairs from

$$\{1, 2, \ldots, n\} \times \{1, 2, \ldots, n\}.$$

Latin Squares have a long history dating to Euler. They also have applications in the design of statistical experiments.

We can derive two 4×4 orthogonal Latin Squares from a 4×4 magic square using the mapping

$$h(a, b) = 4a + b$$

where a and b are entries from the two orthogonal Latin Squares and $4a + b$ is the corresponding entry from the magic square.

In our example the two orthogonal Latin Squares are shown in Table 7.8.

TABLE 7.8: A Pair of
Orthogonal Latin Squares

0 1 2 3	0 1 2 3
1 0 3 2	3 2 1 0
2 3 0 1	1 0 3 2
3 2 1 0	2 3 0 1

The reader should verify that these squares are orthogonal. We can see that the conditions:

$$4a + b = 0 \text{ implies } a = 0, \;\; b = 0$$

$$4a + b = 5 \text{ implies } a = 1, \;\; b = 1$$

$$4a + b = 10 \text{ implies } a = 2, \;\; b = 2$$

and so forth. Hence we have our construction equivalence: weak magic squares to orthogonal Latin Square and conversely.

7.2.3   Sudoku

Another number array game that has gained amazing popularity in the last few years is called *Sudoku*. This puzzle was invented by an American architect, Howard Gams, in 1979 and published by *Dell Magazine* under the name *Number Place*. It became popular in Japan in 1986 where the name Sudoku (a shortening of the phrase for "the numbers must occur once") was used (see [40]). It became an international hit in 2005.

The idea of the game is to fill a 9×9 array with the integers $1, 2, \ldots, 9$ in such a way that each of the nine integers appears once in each row and once in each column (thus making the array a *Latin Square*), and also so that each of nine 3×3 subarrays also contain each of the digits 1 through 9. The player is provided an array with some entries already filled in (see Table 7.9) and must then find a way to properly complete the square.

There are a couple of fundamental strategies people use in trying to complete a Sudoku puzzle. The first might be called *scanning*.

Scanning is usually performed at the outset and periodically throughout the process. There are two fundamental features to scanning. The first is *cross-hatching*. Here you scan the rows to identify which line in a region may contain a given integer. The process is repeated with the columns. If we do this with our example puzzle, we note that the upper right region needs to contain a 5, but the 5 cannot be in the first row as it is in the first row of the upper left-hand region, and it cannot be in the second row as it is there in the upper middle region. Further, it cannot be in the third column as it is there in the lower right region. As there is a 6 in position $(3, 8)$, this leaves only position $(3, 7)$ for the 5.

The second part of scanning is *counting* $1 - 9$ in regions, rows and columns to identify missing integers. For example, the upper left region

TABLE 7.9:   A Partial Sudoku Array

5	3			7				
6			1	9	5			
	9	8					6	
8				6				3
4			8		3			1
7				2				6
	6					2	8	
			4	1	9			5
				8			7	9

is missing 1, 2, 4, and 7. We note the 4 cannot be in column 1, the 7 cannot be in row 1, and the 1 cannot be in row 2. This helps narrow down the possibilities. We hope this will narrow things to two, or at most three possibilities. We have accomplished this with the upper left-hand region.

Scanning stops when we cannot identify any more integers to insert. Now we must undertake some sort of logical analysis of the array.

Some people like to mark cells with a short list of integers possible for that cell. Then later they can compare these lists in the hope of gaining more information.

But finally, at some point, the player must ask the question: What if x appears in location (i, j)? This is usually done only with a cell (i, j) that has only two candidates. The player makes a guess and then tries to complete the array.

Many people have resorted to computer searches to try more options, but this seems to eliminate the "fun of the game."

Beginners may wish to start with smaller arrays. A 6×6 array with 2×3 subregions provides good practice. Here is one such example.

TABLE 7.10:   A 6 × 6 Sudoku Puzzle

		6	2		
4		2			1
	4			5	6
5			4		
	2			1	4
		4		2	

Exercises:

7.2.1 Find a magic square with first row $1, 9, 5$.

7.2.2 Find a magic square with first row $2, 9, 4$.

7.2.3 Are there ways to construct new magic squares from old ones that do not change the magic constant?

7.2.4 Complete the following 4 × 4 magic square:

$$
\begin{array}{cccc}
1 & 14 & 8 & 11 \\
15 & & & \\
12 & & & \\
6 & & &
\end{array}
$$

7.2.5 Use the magic square from the previous exercise to construct two orthogonal Latin Squares of order 4.

7.2.6 Complete the following 5 × 5 magic square:

$$
\begin{array}{ccccc}
11 & 24 & 7 & 20 & 3 \\
4 & & & & \\
17 & & & & \\
10 & & & & \\
23 & & 15 & &
\end{array}
$$

7.2.7 Use the magic square of the previous problem to construct two orthogonal Latin Squares of order 5.

7.2.8 Complete the 6 × 6 Sudoku puzzle from the section.

7.2.9 Complete the Sudoku puzzle from the section.

7.2.10 Complete the following 9×9 Sudoku puzzle.

	9	4	1		5	2		
8	5	4			7	1	3	
1			9					4
2						5		
	5		7		2			
3								1
6			5					7
	4	8	6			3	9	5
	9	1		4	3	6		

7.3 The Tower of Hanoi

The *Tower of Hanoi* puzzle was invented by the French mathematician Edouard Lucas in 1883 (see [43]). The puzzle is often introduced with an elaborate fable of 64 golden disks, each of a different size, piled in order of size, smallest on top to largest on the bottom. The monks of the temple (presumably in Hanoi) are to transfer the pile of disks to a new location. But the disks are fragile and a larger one can never be placed on top of a smaller one. To make matters worse, there is only one intermediate location where the disks may be placed. The legend says that when the monks complete their task, the world will end. Our job is to determine the number of disk moves that must be made in order to properly move the pile of disks and in the process, find a way of accomplishing the task.

In most situations the puzzle is displayed as a pile of disks on one of three pegs; the first peg is the initial position of the disks, the second peg is the intermediate location, and the third peg the final destination (see Figure 7.1). Of course the roles can vary.

In order to solve the puzzle, we will first consider small cases of $n = 1, 2$ or 3 disks and hope we can find a pattern to the solutions. We will assume the pile starts on peg 1, the left most peg, and must end up on peg 3, the right most peg, with peg 2 as the intermediate location. We will also consider the disks as numbered from 1 to n with the smallest being disk 1 and the largest disk n.

With $n = 1$ there is only one move required, take the disk from peg 1 to peg 3. There is not much to learn from this case.

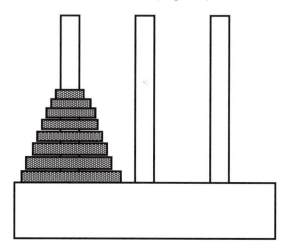

FIGURE 7.1:   The Tower of Hanoi puzzle.

With $n = 2$, the moves required are:

1. Move disk 1 to peg 2.

2. Move disk 2 to peg 3.

3. Move disk 1 to peg 3.

A little thought will convince you that this is the least number of moves possible in this case. What we did here was to move disk 1 to peg 2, the intermediate peg, so that we could move disk 2 to peg 3. Then we completed the moves by placing disk 1 on top of disk 2 on peg 3.

Next we consider the $n = 3$ disk case. The easiest way to think about solving this case is to view it in a manner similar to the $n = 2$ case solution. That is, we must move the top pile of two disks to intermediate peg 2, so that we can move disk 3 to peg 3. We can then complete the process by moving the pile of two disks from peg 2 to peg 3.

Rather than simply counting out all the moves required to complete such a plan, we let M_n be the number of moves required to move a pile of n disks from one peg to another. We already know $M_1 = 1$ and $M_2 = 3$. We can be more complete and assume $M_0 = 0$. We would like to determine M_n for any $n \geq 1$.

Now our process of moving the top pile of two disks, then moving disk 3, then again moving the pile of two disks requires

$$M_3 = M_2 + 1 + M_2 = 2M_2 + 1$$

moves. More generally, our process would be to move the pile of the top $n-1$ disks to peg 2, then move disk n to peg 3, then move the pile of $n-1$ disks from peg 2 to peg 3. Hence,

$$M_n = M_{n-1} + 1 + M_{n-1} = 2M_{n-1} + 1. \tag{7.2}$$

Using this *recursive formula*, also called a *recurrence relation*, (that is, an equation for the n-th value in a sequence of numbers, defined in terms of earlier values of the sequence) we see that

$$M_1 = 1, M_2 = 3, M_3 = 7, M_4 = 15, M_5 = 31, M_6 = 63, \ldots$$

and we could determine M_n for any reasonable n this way.

What we have so far is nice, but can we actually use the recurrence relation to help find a closed form for M_n? Note that in order to "solve" a recurrence relation like (7.2), we need to know some initial condition(s) for the relation. That is, there are infinitely many sequences of values that satisfy the recurrence relation (or almost any recurrence relation), thus, in order to identify a specific sequence we must know some initial values. In our case we know $M_0 = 0$, $M_1 = 1$ and $M_2 = 3$, which is more than enough to allow us to identify the exact sequence.

Returning to Equation (7.2), we define $r_n = M_n + 1$. Then, clearly, as $M_0 = 0$, we have that $r_0 = 1$ and hence for $n \geq 1$,

$$r_n = (2M_{n-1} + 1) + 1 = 2M_{n-1} + 2 = 2(M_{n-1} + 1) = 2r_{n-1}. \tag{7.3}$$

Thus, we have a recurrence relation for r_n defined as $r_n = 2r_{n-1}$, for $n \geq 1$ with the initial condition $r_0 = 1$. But recurrence relation (7.3) implies

$$r_n = 2r_{n-1} = 2(2r_{n-2}) = 2^2(2r_{n-3}) = \ldots = 2^n(r_0) = 2^n(1) = 2^n.$$

Since $r_n = M_n + 1$ for $n \geq 1$, we see that

$$M_n = r_n - 1 = 2^n - 1. \tag{7.4}$$

We can see that our closed form for M_n produces the same values as the recursive formula with the given initial conditions, for all $n \geq 0$.

Thus, we have solved the recurrence relation, producing an easy to use formula.

Finally, we note that even if the monks were to move a disk every second, since there are 64 disks, the transfer would take

$$\frac{2^{64} - 1}{60 \times 60 \times 24 \times 365.2425} \approx 600 \text{ billion years}$$

to complete. This might be considered job security for the monks!

7.3.1   Finding Solutions

Next we ask if there is a method for seeing the sequence of moves we must make? In order to accomplish this we will again call on graphs. This time, in two different ways. Define the *n-cube* Q_n to be the graph on 2^n vertices whose vertices are labeled by the n-tuples from $\{0, 1\}$ such that vertices are adjacent if and only if the corresponding n-tuples differ in only one term (see Figure 7.2).

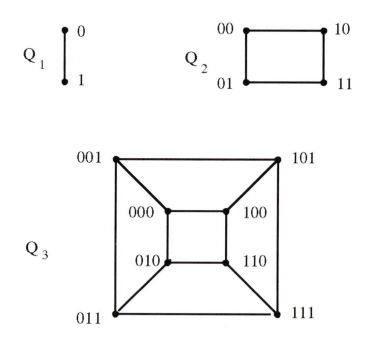

FIGURE 7.2:   The *n*-cubes Q_n, $n = 1, 2, 3$.

The labels on the vertices can be seen to correspond to the moves we shall make. But first we need to find a route through the vertices, starting at the vertex labeled with all zeroes, and eventually returning to that vertex, and visiting all other vertices exactly once. Our moves must be from a vertex to an adjacent vertex, hence from one n-tuple (or vertex label) to another n-tuple that differs in exactly one position. Such a route in the graph that returns to the starting vertex is called a *hamiltonian cycle* of the graph.

In Q_1 the route is obvious (and degenerate as we only have two vertices) and reflects the one move needed in this case. For Q_2, such a hamiltonian cycle is:

$$00, 10, 11, 01, 00. \tag{7.5}$$

The information contained in this cycle is which disk should be moved next. The position in the n-tuple that changes tells us which disk moves. That is, in going from one vertex to the next, the one position that differs in the new label is an indicator of which disk was moved. The number of moves of the disks are recorded *mod* 2; that is, an even number of moves is seen as a 0 and an odd number of moves is seen as 1 in the n-tuple. We should also note that a listing of the n-tuples in this manner constitutes a *Gray Code*, that is, a sequence of n-bit words (the n-tuples composed of 0s and 1s) where the next word differs from the present word in exactly one position. Gray Codes have many important applications.

Following the information from this hamiltonian cycle (Gray Code) we see that:

- The edge from 00 to 10 indicates to move disk 1, as the change is to position 1 of the word.

- The edge from 10 to 11 indicates to move disk 2.

- The edge from 11 to 01 indicates to move disk 1.

The final edge completes the hamiltonian cycle, but is not needed in the puzzle moves (recall $2^n - 1$ moves are sufficient, but there are 2^n edges in the cycle). A solution to the puzzle always allows the completion of the hamiltonian cycle.

In this example we had no choice about the cycle. There are only two possible listings for a hamiltonian cycle in Q_2, that is, really one cycle traversed in opposite directions. But since we clearly must move disk 1 first, we are forced to choose the cycle shown in (7.5). In larger examples, there are even more possible cycles, and the choice of move is not always so clear. What is clear is that if we have a solution to the puzzle, it will allow us to find a hamiltonian cycle (and underlying Gray Code) in the corresponding n-cube Q_n.

There is a second way of using graphs to model the puzzle. Here the vertices of the graph represent the possible disk configurations. For the one disk game the graph is a triangle (see Figure 7.3). The vertices indicate where the one disk can be placed, peg 1, peg 2 or peg 3.

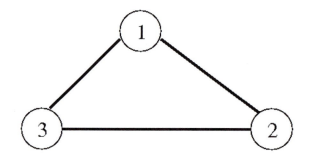

FIGURE 7.3:　Another graph model.

For the graph of the 2-disk puzzle, we arrange 3 triangles, one replacing each vertex of the 1-disk graph (see Figure 7.4). The labels of the vertices again represent the configuration of the disks upon reaching this point in the process. Thus, the label 11 represents both disks being on peg 1, the starting configuration for the 2-disk puzzle. Then 21 represents a move to the configuration of one disk on peg 2 and one disk on peg 1, etc.

To go from the graph representing the 2-disk game to the graph representing the 3-disk game, we again replace each vertex by a triangle (see Figure 7.5). Another way to look at this is to replace each vertex of the triangle from the 1-disk game with a copy of the 2-disk graph and then adjust the labels.

To obtain the new labels in the 3-disk graph we do the following: for the labels in the graph replacing the vertex labeled 1, append a 1 to each label. Thus, that triangle is labeled as shown in Figure 7.5.

For the vertices in the part of the graph replacing the vertex labeled

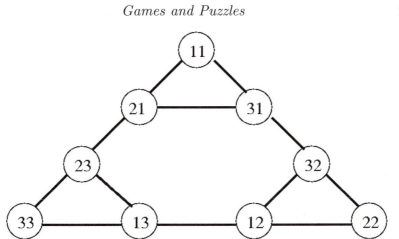

FIGURE 7.4:   The 2-disk graph model.

2, change the old labels by mapping $1 \rightarrow 2$, $2 \rightarrow 3$, and $3 \rightarrow 1$. Now append a 3 to each of the labels.

For the labels in the part of the graph replacing the vertex labeled 3, map the old labels using $1 \rightarrow 3$, $2 \rightarrow 1$, and $3 \rightarrow 2$. Finally, append a 2 at the end of each label. See Figure 7.5 for the final 3-disk graph.

Note that we can read out a shortest solution to the Tower of Hanoi puzzle by following the edges down either side of this triangle. The side of choice depends upon which peg is to be the final peg. Thus, in the 3-disk graph, if peg 3 is the destination, then follow the edges down the right side, while if peg 2 is the destination, then follow the edges down the left side. Thus, the corner vertices can be used to determine which side of the structure you wish to traverse.

We can also use the graph to model the longest nonrepetitive (each vertex of the graph is visited at most once) solution to the Tower puzzle. One such solution for the 3-disk puzzle with peg 2 as the destination is shown in Figure 7.6. Of course, any path from the starting vertex to the destination vertex provides a solution to the puzzle.

7.3.2   Bicolored Tower of Hanoi

There are several variations of the Tower of Hanoi, other than just changing the number of disks. The first is called the *Bicolored Tower of Hanoi*. We shall consider only a simple version of this game.

Again suppose we have three pegs. On the first peg is a pile of n

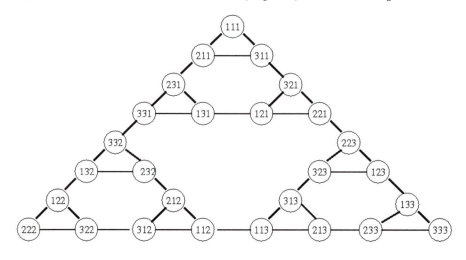

FIGURE 7.5:   The 3-disk graph model.

disks, again increasing in size as before, but also alternately colored black and white. On peg 2 is another pile of n disks, but alternately colored white and black. The object is to move the disks, following the old rule that a bigger disk may never be on top of a smaller disk, so as to obtain two monochromatic piles, that is, one pile of black disks and one pile of white disks.

As an example, we consider two piles of two disks each (see Figure 7.7). Here B1 indicates the smaller black disk, B2 indicates the larger black disk, W1 the smaller white disk, W2 the larger white disk. Then the moves are:

1. Move B1 to peg 3.

2. Move W1 to peg 1.

3. Move B1 to peg 2.

Thus, we obtain two piles, each of a single color, solving the puzzle. Of course the puzzle is more involved with more disks. The 2-disk bicolored game is similar in nature to the ordinary 2-disk Tower of Hanoi. Clearly, other variations of the bicolored Tower of Hanoi are possible, but we will not consider them here. The 3-disk version is considered in the exercises.

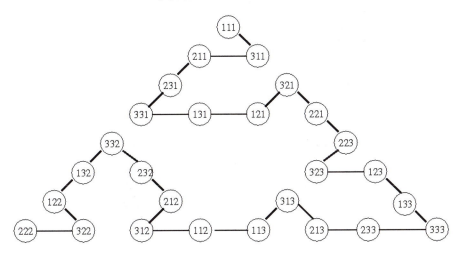

FIGURE 7.6: A longest solution for 3 disks.

7.3.3 The Derangement Tower of Hanoi

The second variation of the Tower of Hanoi puzzle is to start with three piles, one on each peg. Each pile follows the standard rule on disk size and each pile is of a single color, one white, one black and one grey (see Figure 7.8). The object of this version is a derangement of the piles; that is, to move each pile to a new peg. The standard rule of no larger disk placed on a smaller disk still applies. Hence, we will call this the *derangement* version of the Tower of Hanoi.

As a simple example of this variation, consider three piles of one disk each, one white, one black and one grey on pegs 1, 2 and 3, respectively.

Then the moves involved in this example are:

1. Move the black disk to peg 3.

2. Move the white disk to peg 2.

3. Move the black disk to peg 2.

4. Move the grey disk to peg 1.

5. Move the black disk to peg 3.

We end with the grey disk on peg 1, the white disk on peg 2 and the black disk on peg 3. This completes a shortest solution to the 1-disk derangement puzzle. Larger versions are considered in the exercises.

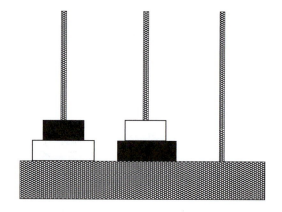

FIGURE 7.7:   A 2-disk bicolored Tower of Hanoi.

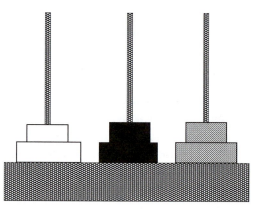

FIGURE 7.8:   A 2-disk derangement Tower of Hanoi.

Exercises:

7.3.1 How many moves are involved in a longest nonrepetitive solution to the standard Tower of Hanoi puzzle with n disks?

7.3.2 Is there a nonrepetitive solution for the Tower of Hanoi puzzle for every number of moves from $2^n - 1$ to the longest number of moves (the value found for the previous exercise)?

7.3.3 Find a 5-move solution to the 2-disk Tower of Hanoi puzzle.

7.3.4 Find a 6-move solution to the 2-disk Tower of Hanoi puzzle.

7.3.5 Find a 10-move solution to the 3-disk Tower of Hanoi puzzle.

7.3.6 Find a solution to the 3-disk bicolored Tower of Hanoi puzzle. How many moves were involved? Is this number optimal?

7.3.7 Find a solution to the 2-disk derangement version of the Tower of Hanoi; that is, with 2 black disks on peg 1, 2 white disks on peg 2 and 2 blue disks on peg 3.

7.3.8 Find a solution to the 2-disk bicolored Tower of Hanoi puzzle with the added condition that, at the end, both largest disks are moved to new pegs.

7.4 Instant Insanity

The game of "Instant Insanity" is a puzzle consisting of four cubes with faces colored from a set of four colors (say red, white, blue, and green). The distribution of colors on each cube is unique (although variations of the game could allow repetitions). The object of the game is to stack the cubes in a $1 \times 1 \times 4$ column so that each side of the column shows each of the four colors.

Credit for inventing the puzzle goes to Franz Owen Armbruster (also called Frank Armbruster) (see [19]), but the puzzle has similarities to an older puzzle known as "The Great Tantalizer." The game was marketed by Parker Brothers beginning in 1967.

Of interest to us is another use of graphs in finding a solution to this puzzle. In order to do this we first create a standard representation for a cube (see Figure 7.9), showing the colors of each of the four sides. Of course, the choice of the front for each cube is completely up to you. Note however that, once you choose the front, then the back is fixed. Then once you select the left, the right is fixed, as are the top and bottom. So you can always obtain this standard representation for each cube.

Next, using our standard cube representation, suppose we are given the four cubes as shown in Figure 7.10.

Using the cubes we create four graphs, one to represent each of the cubes. Each of these graphs consists of four vertices, one labeled for each color. We then draw an edge between colored vertices representing opposite faces; that is, we connect the vertices representing the colors of the top and bottom, left and right, and front and back of the cube.

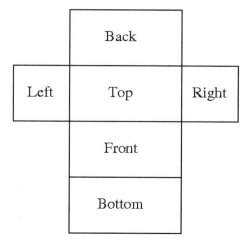

FIGURE 7.9:   The standard representation for a cube.

Hence each graph has four vertices and three edges. Note that an edge can be a *loop*; that is, an edge from a vertex to itself. This happens in the graph representing cube 1, as the top and bottom of cube 1 are both white. Corresponding to the cubes $C_1, \ldots, C_4$, we have the graphs $G_1, \ldots, G_4$ shown in Figure 7.11.

The next step is to superimpose the graphs $G_1, \ldots, G_4$ onto a single set of four vertices (again labeled with the four colors), labeling each edge with the number of the cube that edge represents. Call this superimposed graph S (see Figure 7.12).

Now suppose a solution to the puzzle exists; that is, we can properly stack the cubes so the four colors appear on each side. Consider the front of the stack first and at the same time the back. Since each color is to appear on the front (and back), if a solution exists, there must be a subgraph S_1 of S which represents the front and back of the stack for each of the cubes. That is, there must be a subgraph S_1 which contains four edges, with each edge labeled by a different cube number, and where each edge joins the front to the back of one of the cubes. Hence, each of the four vertices (colors) should have two edges incident to it, one from the color appearing on the front of the column and one for the color appearing on the back of the column.

Thus, each vertex of S_1 has degree 2 (loops count two towards degree) since each color appears once in the front and once in the back of the stack. Hence, S_1 is what is called a 2-*regular graph* (that is, each

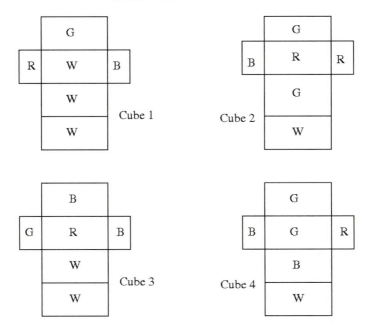

FIGURE 7.10: The four cubes.

vertex has degree two, or in general r-regular where each vertex has the same degree r). Hence, our goal is to find two edge disjoint 2-regular subgraphs S_1 and S_2 of S. Then S_1 can be used to determine the front and back of the column and S_2 can be used to determine the left and right sides of the column. In our example these two subgraphs are shown in Figure 7.13.

With these two subgraphs we are now able to determine a way to stack the cubes properly. We start with cube 1 and place its red side in front and its blue side in back (using the red to blue edge labeled 1 in S_1). In effect, we have determined a direction on that edge from the front (red) to the back (blue). This direction is inherited by all edges along the cycle in S_1. At the same time, we also place white on the right of cube 1 and green on its left, using the edge labeled 1 from S_2. Again we have assigned an orientation to the edge and this will be inherited by all edges along the cycle in S_2. If it helps, you could assign these directions and then follow them in placing the cubes in the column. Of course we could have oriented it the opposite way without causing any problems at this stage.

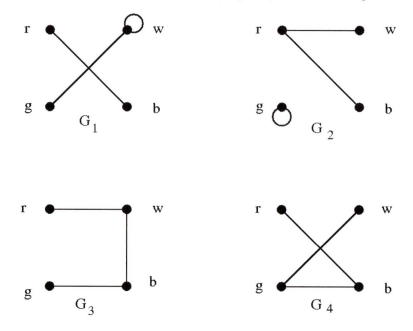

FIGURE 7.11:   The corresponding graphs.

Next we consider cube 3, as its edge comes next following the orientation of the cycle in S_1. We place blue in front and green in back; white on the left and red on the right, based on the earlier choices and the adjacencies in S_1 and S_2. Now cube 2, where we place white in front and red in back while red appears on the left and blue on the right. Finally we consider cube 4, where we place green in front and white in the back with blue on the left and green on the right. This stack now solves the puzzle.

Thus, to recap, there is a set algorithm that will always lead to a solution of the puzzle when one exists. The steps in solving Instant Insanity are:

- Given the four blocks, randomly call them cube 1, cube 2, etc.

- Create the standard representation for each cube.

- Create the graph representation for each cube.

- Create the superimposed graph S.

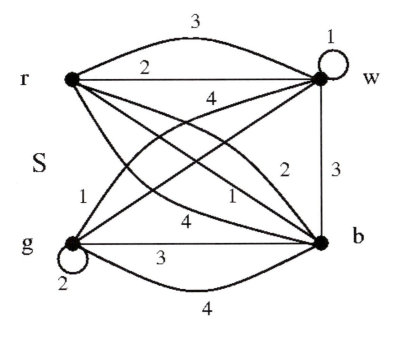

FIGURE 7.12: The superimposed graph S.

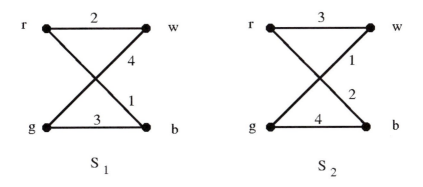

FIGURE 7.13: The two subgraphs S_1, S_2.

- Find the two 2-regular edge disjoint subgraphs of S. If these two subgraphs do not exist, then there is not a solution for the given cubes and this option is certainly possible.

- Given the two 2-regular graphs from the previous step, use them to determine how to build the stack. Remember to impose a direction in order to maintain front to back type orientation properly.

This algorithm will always allow you to solve the puzzle.

Exercises:

7.4.1 Determine if there exists a solution to Instant Insanity with the family of cubes shown in Figure 7.14.

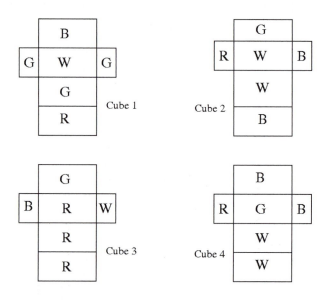

FIGURE 7.14:   The cubes of Exercise 7.4.1.

7.4.2 Find a solution to Instant Insanity with the cubes of Figure 7.15.

7.4.3 Find a solution to Instant Insanity with the cubes of Figure 7.16.

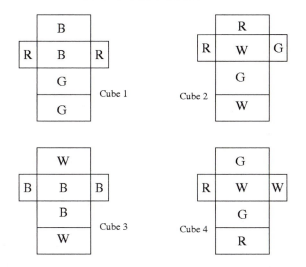

FIGURE 7.15:   The cubes of Exercise 7.4.2.

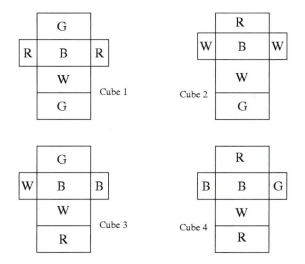

FIGURE 7.16:   The cubes of Exercise 7.4.3.

7.4.4 Find a set of cubes, each colored with all four colors, where there is no solution.

7.4.5 Is it necessary for all four colors to appear on each cube? Explain your answer.

7.4.6 Is it necessary that at least two colors appear on each cube?

Explain your answer.

7.4.7 How many arrangements of the four cubes are possible? (Hint: the answer is $41,472$; now show how this count occurs.)

7.4.8 What is the maximum number of solutions possible from the $41,472$ arrangements?

7.4.9 Show that the superimposed graph of the example from this section does not contain three edge disjoint 2-regular subgraphs.

7.5   Lights Out

Lights Out is another electronic puzzle, first released by Tiger Toys in 1995 (Tiger Toys was later purchased by Hasbro in 1998). The game consists of a 5×5 grid of lights. At the start of the game a random pattern of lights are turned on. Pressing a light changes the state of the light; that is, on to off or off to on. Pressing a light also changes the state of the horizontal and vertical neighbors (if they exist) of the light. Note that diagonal neighbors are not affected. The goal is to turn out all the lights, hence the name of the game!

Note that Parker Brothers (1970s) had a mini version of the game called *Merlin* that was played on a 3×3 grid. Some other companies also produced similar games. A 4×4 version will be discussed later. For simplicity we will discuss a 2×2 version, as the principles remain the same for the larger versions.

There are two important points one needs to note when considering Lights Out games of any size.

1. Each light needs to be pressed no more than once. Pressing a light twice returns it to its original state, which is equivalent to not pressing it at all.

2. The order in which you press the lights does not matter. The result of pressing a given set of lights is always that each light has been changed a certain number of times dependent upon the location of that light to those lights pressed. This count does not change by pressing the lights in a different order, since the final count remains unchanged.

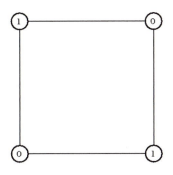

FIGURE 7.17:   A 2×2 Lights Out grid graph.

Now consider the game shown in Table 7.11. Here we use an array to model the light grid, with a 1 to indicate the light is on and a 0 that the light is out. The array (matrix) models the game which can also be modeled by the grid graph shown in Figure 7.17.

TABLE 7.11:   Lights Out Game

1	0
0	1

Consider the actions taken by pressing a light. The set of lights that change state is clearly dependent upon which light is pressed. A light can be in a corner of the grid, along the edge of the grid, or (for larger games) within the center of the grid. Each produces a different pattern of light changes. We can model these moves by arrays as well. Typical corner, edge and center moves for a 3×3 Lights Out game are modeled in the arrays of Table 7.12. Other size games have similar move arrays.

Perhaps the single most important point to take note of is that the solution to the puzzle is equivalent to the set of moves that would turn the present set of lights on. If we can find the set of lights that would turn the present lights on (from a starting grid where all lights were out), then pressing these lights will reverse the process and solve the puzzle. This is useful because we know the set of lights that are on and we know for each light how pressing that light affects the set of lights.

TABLE 7.12:   Typical Corner, Edge and Center Move Arrays

$$\begin{vmatrix} 1 & 1 & 0 \\ 1 & 0 & 0 \\ 0 & 0 & 0 \end{vmatrix} \quad \begin{vmatrix} 1 & 1 & 1 \\ 0 & 1 & 0 \\ 0 & 0 & 0 \end{vmatrix} \quad \begin{vmatrix} 0 & 1 & 0 \\ 1 & 1 & 1 \\ 0 & 1 & 0 \end{vmatrix}$$

We can then model turning these lights on with a system of equations.

Label the lights $x_{i,j}$ where i, j refers to row i and column j of the grid. Now we consider the move arrays for each light. In our example 2×2 game we obtain the following equations.

Considering light $x_{1,1}$ and its move array we see that

$$x_{1,1} + x_{1,2} + x_{2,1} = 1.$$

Considering light $x_{1,2}$ we see that

$$x_{1,1} + x_{1,2} + x_{2,2} = 0.$$

Considering light $x_{2,1}$ we see that

$$x_{1,1} + x_{2,1} + x_{2,2} = 0.$$

Finally, considering light $x_{2,2}$ we see that

$$x_{1,2} + x_{2,1} + x_{2,2} = 1.$$

Adding the first and second equations and recalling that pressing a light two times is equivalent to zero times we see that $x_{2,1} + x_{2,2} = 1$.

Using this fact in the fourth equation implies that $x_{1,2} = 0$. Now adding equations one and three implies that $x_{1,2} + x_{2,2} = 1$. This implies that $x_{2,2} = 1$. Now substituting into equation two we see that $x_{1,1} = 1$ and finally from equation three we see that $x_{2,1} = 0$.

This system of four equations in four unknowns has the solution $x_{1,1} = 1$ and $x_{2,2} = 1$ while $x_{1,2} = x_{2,1} = 0$. Thus, we can shut off all the lights by pressing the $(1, 1)$ light and the $(2, 2)$ light.

As mentioned earlier, this is the approach to any sized grid. However, for larger systems it is useful to use the tools of linear algebra, which are beyond the scope of this book. It was shown by Sutner [41] that

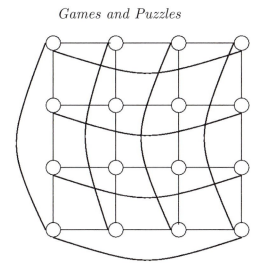

FIGURE 7.18:   Mini Lights Out model.

there is a solution to any light pattern on any square grid. Anderson and Feil [3] used standard linear algebra to solve the 5×5 grid while Missigman and Weida [25] solved the 4×4 Mini Lights Out version.

Earlier we mentioned that there is a 4×4 version of the game as well. This version is sometimes called *Mini Lights Out*. There is one fundamental difference in Mini Lights Out. That is the underlying graph is more than just a grid. Each light along the top edge is considered a neighbor of the corresponding light along the bottom edge. Similarly, each light along the left edge is considered a neighbor of the corresponding light along the right edge. See Figure 7.18 for a graph model of this game.

Exercises:

7.5.1 Solve the 2×2 Lights Out Game:

1	1
0	0

7.5.2 Solve the 2×2 Lights Out Game:

1	1
1	1

7.5.3 Solve the 2×2 Lights Out Game:

$$\begin{array}{|c|c|} \hline 1 & 1 \\ \hline 0 & 1 \\ \hline \end{array}$$

In the following questions we play lights out on a board modeled by some graph. We will use the same rule, pressing any light changes that light and all its neighboring lights (those connected by an edge).

7.5.4 Suppose we play lights out on the graph that is a path with three vertices. Suppose the only light that is on is at one end vertex. Argue why this game cannot be won; that is, we can never shut off all the lights. (Hint: use the fact it is useless to repush a light.)

7.5.5 Find a lights out game on a path with three vertices that can be won.

7.5.6 Suppose we play lights out on the graph that is a path with four vertices. Suppose the only light that is on is at one end of the path. Argue why we can never win this game. Would the result change if the path was longer?

7.5.7 Suppose we play lights out on the complete graph on five vertices (that is, each vertex has an edge to every other vertex). Suppose only one light is on. Argue why we cannot win this game.

7.5.8 Suppose we play lights out on a complete graph with five vertices minus one edge, say the edge from vertex x to vertex y. Suppose light x is the only one on. Can we win this game?

7.5.9 Show that you can win a lights out game on a 5-cycle with one light on.

7.6 Peg Games

Another old and well-known game is *Peg Solitaire*. There are a number of versions presently available, but we will concentrate on only two:

the English board version and the triangle (made famous in Cracker Barrel© restaurants) version. Each of the games have fundamentally the same rules, it is the playing board that varies and makes them different.

Each game is played on a board with some number of holes. At the start of a game, usually all but one hole is filled with a peg. A peg may jump over a neighboring peg. The jumped peg is then removed from the board. (We shall see that which pegs are considered neighboring pegs is determined by the board in use and the version being played.) The object is to reduce to only one peg, preferably with this peg in the original open hole.

7.6.1 English Board

We begin with peg solitaire on an English board, marketed under a variety of names including Hi-Q by Milton Bradley, 1967. The game draws its name directly from the board used. It is played on a board with 33 holes in which 32 of the holes initially have a peg inserted. The configuration of the holes is shown in Figure 7.19. Black circles are filled holes, the open circles are unfilled holes. The usual game begins with only the center hole empty, called the *central game*. A peg may jump over a neighboring peg in the horizontal or vertical directions.

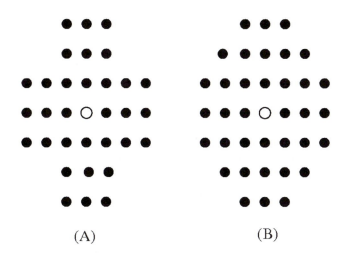

(A)   (B)

FIGURE 7.19: English board (A) and European board (B).

The English board differs slightly from the European board which has four more holes (and hence pegs) located in the four corners of the cross (see Figure 7.19). Legend states that the game was originally invented by French nobles in the 17th century, while they were imprisoned in the Bastille. However, to date there is no real evidence to confirm this tale. The first solid evidence of the game dates to the court of Louis XIV, circa 1697, where several works of art from that time period show peg boards, implying the game was fashionable (see [34]) among the French nobility.

For convenience of description we shall label the holes with the lattice points of a standard $X - Y$ coordinate system, using the center hole for $(0,0)$. We shall make use of some other fundamental mathematics to gain insight into this game. In particular, where the final peg may reside. To accomplish this we define the following addition table for the set of four elements 0, x, y and z.

TABLE 7.13:	Addition Table for the Special Group of 4 Elements

+	0	x	y	z
0	0	x	y	z
x	x	0	z	y
y	y	z	0	x
z	z	y	x	0

Under these rules we see that 0 is the *additive indentity* element since $0 + w = w$ for $w = 0, x, y$, or z. Each element also has the interesting feature of being its own *additive inverse*, that is,

$$x + x = y + y = z + z = 0.$$

Another interesting fact about this system is that the sum of any two distinct nonzero elements is equal to the third nonzero element. That is, $x + y = z$, and $x + z = y$, and $y + z = x$. We also note and can easily verify from the addition table, that this special addition is *associative*;

that is, for all a, b, c in this system,

$$(a + b) + c = a + (b + c).$$

Further, it is also *commutative*; that is, for all a, b in this system,

$$a + b = b + a.$$

We use the nonzero elements of this system (for those of you who know a bit more, this is known as the Klein 4 group) to label the holes on the English board. Such a labeling is shown in Figure 7.20. Using this labeling and the fact that

$$x + y + z = x + y + x + y = 0,$$

we see that adding all the values of the labels on the holes of the board equals zero.

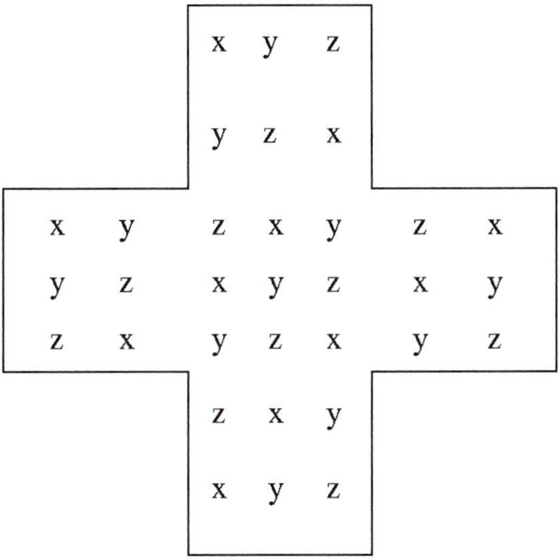

FIGURE 7.20: Labeling the holes.

Now given any configuration C of pegs on the board, we define the value of this configuration as $V(C)$, where $V(C)$ is found by adding the labels of all the filled holes corresponding to this configuration. Under this definition we note that the value of the initial configuration, with only the center hole empty, must be $V(C) = y$. Further, the key observation is that any legal move does not change the value of V, even though the configuration does change. This follows because for any legal move, the sum of two elements from $\{x, y, z\}$ is replaced by the third element. But we know that the sum of any two of these elements equals the third!

Consequently, every conceivable configuration of pegs falls into one of four classes which correspond to the $V(C)$ values 0, x, y or z. But since our game begins with $V(C) = y$, and any legal move does not change the value of V, the ending value of V must also be y. That is, the final peg must be in a hole labeled y! But we can say even more. The following result is due to A. Bialostocki [9].

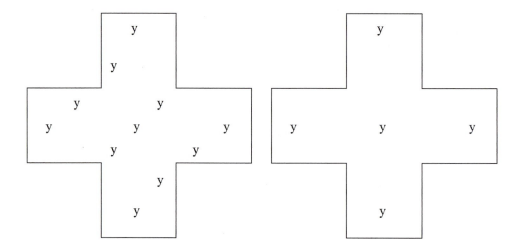

FIGURE 7.21:   Eleven holes labeled y and the five that are possible solutions.

Theorem 7.6.1. *There are at most five locations in which a single peg can form the final configuration of the English board peg game, namely*

$$(0, 0), \quad (0, 3), \quad (0, -3), \quad (3, 0), \quad and \ (-3, 0).$$

Proof: Since $V(C) = y$ at the outset of the game and the value of V can not change during the game, the final peg can be in only one of eleven holes labeled y (see Figure 7.20 and Figure 7.21).

However, we note that by the symmetry of the board, if the final peg is in location $(1, 1)$ (which is labeled y), then it is also possible to leave the final peg at $(1, -1)$ which is labeled z. (Just think of reflecting the board around the Y-axis and repeating the same moves.) But this contradicts the fact that V cannot change value during the game. Hence, we conclude we cannot end the game with only one peg at position $(1, 1)$. A similar argument eliminates all positions of the 11 labeled y except (see Figure 7.21)

$$(0, 0), \quad (0, 3). \quad (0, -3), \quad (3, 0), \text{ and } (-3, 0).$$

□

We next turn our attention to finding a solution. In attempting to "solve" the puzzle, it will be helpful to be able to know the effect of a series of moves, before they are made. We follow [7] and call such series *packages*.

When you look at the English board you soon realize that in order to solve the central game (last peg in the center hole), you must clear the top, bottom, left and right regions of the board in an effective manner. This leads us to the idea of a 6-*purge*, see Figure 7.22, where the two Xs represent one filled and one unfilled hole. The object of the 6-purge is to clear the six pegs in the region and return the filled X to its original position. To accomplish a 6-purge, we first use a 2-package and then a 4-package (see Figure 7.22).

There are several other such series that are very useful. We next show the 3-purge and L-purge (see Figure 7.23).

Problem 7.6.1. *Write out the series of moves for:*

- *A 6-purge.*

- *An L-purge.*

With the aid of these basic packages we can write down a solution to the central peg puzzle. This solution is presented elegantly in Figure 7.24. Here there are two 3-purges (packages 1 and 2), followed by three 6-purges (packages 3, 4 and 5), followed by an L-purge (package 6) leaving only the final jump to be made. We perform the purges in numerical order.

Problem 7.6.2. *Write out the series of moves involved in the solution to the central puzzle shown in Figure 7.24.*

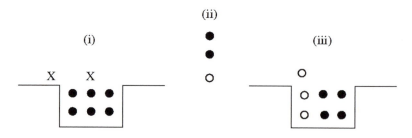

FIGURE 7.22:   A 6-purge (i), 2-package (ii) and 4-package (iii).

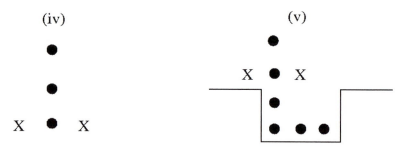

FIGURE 7.23:   A 3-purge (iv) and an L-purge (v).

7.6.2   Triangular Peg Solitaire

Another well-known peg solitaire version is played on a 15-hole triangular board, with 14 holes filled with pegs. The neighboring pegs are determined by the edges of the graph model shown in Figure 7.26. The claim to fame for this puzzle is that they can be found at the tables of Cracker Barrel restaurants. Another feature is that this puzzle is amenable to exhaustive computer search, so many computer science students have programmed solutions over the years.

To discuss the triangular board and moves, it is convenient to use a different system to recognize position. Namely, we will just attach an alphabetic character to each hole as shown in Figure 7.25. Neighboring pegs are determined by the edges of the graph model shown in Figure 7.26.

We first ask if we can determine the possible end positions of a single peg, as we did with the English board. It is natural to ask if a labeling

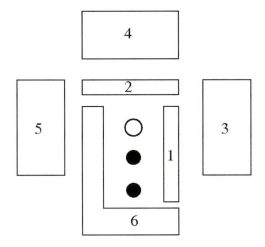

FIGURE 7.24: A package solution to the central puzzle.

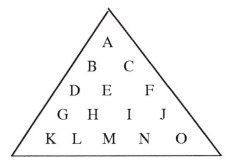

FIGURE 7.25: A labeled triangular board.

similar to that of the English board can be found? Luckily, the answer is yes! This labeling works as it did before, in that for any move, the sum of two labels converts to the third label. Thus, the value of the board in the standard initial position of all holes filled except the top of the triangle will have value x. Again the value never changes and so the only possible one-peg solutions for the standard starting position are at holes labeled x. We see there are five such holes and none are excluded by symmetry.

Theorem 7.6.2. *The standard triangular game can end with one peg in one of the holes labeled with value x, namely in one of holes A, E, G, J, or N.*

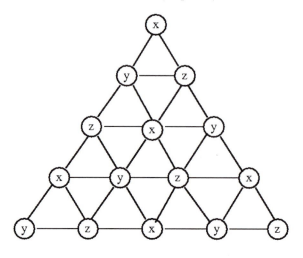

FIGURE 7.26:   Labeling of the triangular board.

Unfortunately, the packages that were so useful for the English board are not so helpful here, since this board differs considerably in shape. But we will demonstrate one solution to the puzzle. Here *X-Y* means jump from hole X to hole Y. The initial missing peg will be at hole A in this example.

A solution for the 15-Triangular Board:

$$D - A, \ F - D, \ L - E, \ N - L, \ K - M, \ O - F, \ C - J, \ M - F,$$

then

$$G - B, \ J - C,$$

followed by

$$A - D, \ D - F, \ F - A.$$

Note there are 13 moves for the 14 pegs, as small as is possible.

Exercises:

7.6.1 Find another solution to the central game on the English board. Are there still other solutions that are easy to find?

7.6.2 If we begin a game on the English board with only hole $(0, -1)$ unfilled, then what are possible positions for a final single peg? Can this list be reduced?

7.6.3 If we begin a game on an English board with only position $(1, 2)$, then what are the possible positions for a final single peg? Can this list be reduced?

7.6.4 Suppose we start a peg solitaire game on an English board with two open holes. Speculate on where the final peg must be if there is a one-peg solution and the two open holes are labeled x and y. What if the two open holes are both labeled z?

7.6.5 If we begin a triangular board game with only the lower left corner hole unfilled, then what are the possible positions for a final single peg? Can this list be reduced?

7.6.6 If we begin a triangular board game with only the lower right corner hole unfilled, then what are the possible positions for a final single peg? Can this list be reduced?

7.6.7 Find a solution for the 15-peg triangular board with the lower left corner unfilled at the start.

7.6.8 Find a solution for the 15-peg triangular board with the lower right corner unfilled at the start.

7.6.9 Complete these moves for a game on a 15-peg triangular board with the upper hole unfilled.

$$D - A, \ M - D, \ K - M, \ N - L, \ F - M, \ L - N, \ O - M.$$

7.6.10 Write out the moves in the solution to the English board game shown in Figure 7.27, where the open hole is at location $(-1, 1)$.

7.6.11 Write out the moves in the solution to the English board game shown in Figure 7.28, where the open hole is at location $(0, 1)$.

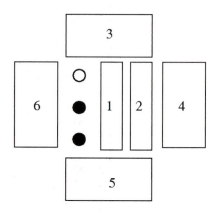

FIGURE 7.27:   Solution to the English board with $(-1, 1)$ open.

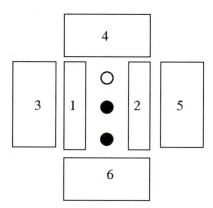

FIGURE 7.28:   Solution to the English board with $(0, 1)$ open.

Chapter 8

Combinatorial Games

8.1 Introduction to Combinatorial Games

Combinatorial games are a special class of games that have received a great deal of attention (for example, see [6], [7], [30], [12]). A vast general theory has been developed, far more than we can cover here. Thus, our goals shall be a bit less lofty. Instead, we shall concentrate on a number of fundamental games to see what we can learn directly from each game. A tiny bit of the general theory will slip in along the way.

In general, combinatorial games satisfy the following special conditions:

- There are just two players (often designated as Left and Right). There can be no coalitions.

- There are usually finitely many *positions* possible and often there is a particular starting position for the game.

- There are clearly defined rules that specify the two sets of moves that Left and Right can make from a given position.

- Left and Right move alternately. Who moves first may vary from game to game.

- The convention is a player unable to move loses. This is called the *terminal position*.

- The rules are such that play will end because some player is unable to move, so no draws are possible.

- Both players know what is happening; that is, there is *complete information* and hence, no bluffing.

- There are no chance moves, so no dealing of cards or rolling dice.

As we shall see, there are many, many games that fall into this class. We will begin with a very simple game that will eventually lead us to the heart of combinatorial games.

8.2   Subtraction Games

An *impartial game* is one in which the set of moves available from any given position is the same for both players. As an example of a simple impartial combinatorial game, consider the following *subtraction game*.

There are two players: player I and player II. There is a pile of 17 chips. A move consists of one player removing one, two or three chips from the pile. Players alternate moves, starting with player I. The first player with no move loses.

This simple game can be played in many ways. But is there a strategy for playing this game? To see this, we will analyze the $n = 17$ game from the end back to the beginning.

Clearly, if there are one, two or three chips left, the player who moves next will win by taking all the chips. If there are four chips left, then the player who moves next is forced to leave one, two or three chips on the table and thus will lose on the next move. If there are 5, 6 or 7 chips on the pile, then the next player can reduce the pile to four chips and ensure victory. With 8 chips left, the next player must leave 5, 6 or 7 chips in the pile and will lose.

Continuing with this argument we see that when there are 0, 4, 8, 12 or 16 chips in the pile, the next player is doomed to lose. We say these are winning positions for the previous player or a **P**-position. On the other hand, 1, 2, 3, 5, 6, 7, 9, 10, 11, 13, 14, 15 and 17 are winning positions for the next player and are called **N**-positions. So in this game, **P**-positions are just those that are a multiple of 4 while **N**-positions are all other positions. In impartial combinatorial games one can find (at least in principle) which positions are **P**-positions and which are **N**-positions.

It is easy to see that the winning strategy is moving to the **P**-positions. Also, from a **P**-position your opponent can only move to an **N**-position. Then you may again return the game to a **P**-position. Eventually the game ends at the **P**-position of zero chips left and you as the winner.

This simple subtraction game is the tip of the iceberg of possibilities. The pile of chips can be of any size. The rule for how many chips you can remove can be any set of possibilities. This set of moves is called the *subtraction set*.

Exercises:

8.2.1 Suppose you have a large pile of chips and your subtraction set is $S = \{1, 2, 3, 4, 5\}$. Determine the proper strategy for this game.

8.2.2 If in the previous exercise there are 41 chips in the pile, then what are the **P**-positions and what are the **N**-positions?

8.2.3 If there are 21 chips in the pile and $S = \{2, 4, 6, 8, 10\}$, then what are the **P**-positions and the **N**-positions?

8.2.4 If there are 33 chips in the pile, determine the set of **P**-positions when the subtraction set is:
(1) $S = \{1, 2, 5, 7\}$.
(2) $S = \{1, 3, 5\}$.
(3) $S = \{2, 4, 6, 8, \ldots\}$.

8.2.5 (The SOS Game: from the 28th USA Mathematical Olympiad, 1999.) The board consists of a row of n squares, initially empty. Players take turns selecting an empty square and writing either an S or an O in the square. The player who first completes SOS in consecutive squares wins the game. If the board fills without SOS appearing, it is a tie.
(1) If $n = 4$ and the first player puts an S in the first square, show that the second player can win.
(2) If $n = 7$, determine if the first player can win.

8.3   Nim

The most fundamental combinatorial game is another simple subtraction game called *Nim*. The game was completely analyzed by Professor C.L. Boulton from Harvard University in 1902 [10]. However, the game is believed to be much older. Only much later was the central importance of Nim in Combinatorial Game Theory realized. We shall see that importance later.

Nim has caught the attention of many people over the years. Martin Gardner [15] mentions a number of mechanical devices built in the 1940s and 1950s that played a perfect Nim game. The most notable of these was called *Nimatron*, built in 1940 and weighing one ton!

The game of Nim is played with piles of chips (or heaps of matches, pebbles, etc.). Players alternately make moves by removing a positive number (at least one and possibly all) of the chips from exactly one of the piles. The first player who cannot move loses. Thus, the rules are truly simple. The game has infinitely many starting positions, as there can be any number of piles of chips and there can be any number of chips in each pile. The theory of how to play works no matter what the position.

We begin by asking a couple of basic questions about the game.

Question 8.3.1. *Can we determine who will win this game in advance (assuming expert play, this is the full information assumption)?*

Question 8.3.2. *Can we determine what play is actually the best in any given position?*

Before we attack these questions, we need to define some notation. Suppose we are at a position in a Nim game where there are k piles with p_i chips in each pile ($i = 1, 2, \ldots, k$). We denote this position as

$$< p_1, p_2, \ldots, p_k > .$$

In order to describe play in a game, consider the following example. Suppose player 1 is in position $< 1, 1, 2, 4 >$ and for her turn she removes three chips from the fourth pile, then we will describe that move as

$$< 1, 1, 2, 4 >_1 \rightarrow < 1, 1, 2, 1 >_2 .$$

That is, the first position was that confronting player 1, and the resulting position is that confronting player 2.

Example 8.3.1. *Suppose we are playing a Nim game with three piles with 1, 1, and 2 chips, respectively. Who will win?*

Solution: The winner will depend upon who moves first. Suppose for this game player 1 must move first. Also suppose that player 1 removes one chip from the third pile, that is $< 1, 1, 2 >_1 \rightarrow < 1, 1, 1 >_2$. From this point on there are no real choices. The play is equivalent to

$$< 1, 1, 1 >_2 \rightarrow < 1, 1, 0 >_1 \rightarrow < 1, 0, 0 >_2 \rightarrow < 0, 0, 0 >_1.$$

Thus, player 1 is confronted with the empty board and has no moves, and so player 1 loses.

But this approach actually represents poor play. The first move by player 1 actually determined all the rest, but this play was a mistake. If instead, player 1 had removed all the chips from pile three, the game would have proceeded as:

$$< 1, 1, 2 >_1 \rightarrow < 1, 1, 0 >_2 \rightarrow < 1, 0, 0 >_1 \rightarrow < 0, 0, 0 >_2.$$

Again the game had no real choices after the first move, but now player 2 has no move and loses. Thus, this play is the superior one for player 1. □

From this point on we will assume each player knows exactly the best move available at the time.

It is here that we should note an important fact. In the previous example, once player 1 had removed all the chips from pile three, reducing the game to $< 1, 1, 0 >$, then player 1 had ensured victory. Player 2 had to remove a chip, leaving only one pile left which player 1 would completely remove, no matter how large it was. This observation is generalized in the following lemma.

Lemma 8.3.1. *For any positive integer r, the Nim game $< r, r >_i$, $i = 1$ or 2, leads to a loss for player i.*

Solution: There are only two nonempty piles of chips, each with the same number r of chips ($r > 0$). Player i must remove some number of chips, say x, from one pile. Then player j merely copies this move, taking x chips from the other pile. The game is now $< r-x, r-x >_i$ and player i is confronted with essentially the same problem. Eventually player i must remove the last chip from one of the piles, and player j will do the same for the other pile, leaving player i facing $< 0, 0 >_i$, and hence a loss. □

Now we make our next important observation. The game ends when there are no piles of chips left. Assigning a value to this terminal position, we can think of this as a zero position.

It is at this point that Boulton made his observation on how to assign values to all other positions. Our ordinary number system is a base 10 system. In theory, we could use any other base. However, in order to understand Boulton's strategy for Nim, we must first consider base two numbers. Later it will also serve us well to understand other number bases, in addition to base 2. We begin with the following general result.

Theorem 8.3.1. *Let b be a positive integer greater than 1. Then if m is a positive integer, it can be expressed uniquely in the form*

$$m = a_k b^k + a_{k-1} b^{k-1} + \ldots + a_1 b + a_0$$

where k is a nonnegative integer and $a_0, a_1, \ldots, a_k$ are nonnegative integers less than b and $a_k \neq 0$.

Our first interest is when $b = 2$ and hence the a_i's are either 0 or 1. When the base is 2 we sometimes say we are finding the *binary* representation for the number. For clarity, in the remainder of this section, we will use subscripts to tell you what base we are using when there could be some question.

Example 8.3.2. *Represent 37_{10} as a base 2 number and then as a base 3 number.*

Solution: We need to represent 37_{10} as the sum of powers of 2. That is,

$$37_{10} = 1 \times 2^5 + 0 \times 2^4 + 0 \times 2^3 + 1 \times 2^2 + 0 \times 2 + 1 \times 2^0,$$

hence, $37_{10} = 100101_2$.

Now as a base 3 number we have:

$$37_{10} = 1 \times 3^3 + 1 \times 3^2 + 0 \times 3 + 1 \times 3^0,$$

hence, $37_{10} = 1101_3$. □

We note there is an easy algorithm for finding the base 2 (or base b) representation for a base 10 number. We exhibited this algorithm on the example above.

For a base 2 representation, first divide 37 by 2 yielding $37 = 2 \times 18 + 1$. Now successively divide the quotients by 2, that is:

$$37 = 2 \times 18 + 1$$
$$18 = 2 \times 9 + 0$$
$$9 = 2 \times 4 + 1$$
$$4 = 2 \times 2 + 0$$
$$2 = 2 \times 1 + 0$$
$$1 = 2 \times 0 + 1.$$

Now, reading the remainders from bottom to top we have our representation, namely $37_{10} = 100101_2$.

Similarly, expressing 37_{10} as a base 3 number we obtain:

$$37 = 3 \times 12 + 1$$
$$12 = 3 \times 4 + 0$$
$$4 = 3 \times 1 + 1$$
$$1 = 3 \times 0 + 1.$$

Hence, reading the remainders from bottom to top we obtain the base 3 representation $37_{10} = 1101_3$.

We can do arithmetic with base 2 (or base b) numbers as well. Addition is done exactly as in base 10, except we must stay within the allowed digits for the particular base. Note the carrying of digits in the sums below.

For example:

$$
\begin{array}{r}
101011_2 \\
+ \quad 110010_2 \\
\hline
1011101_2
\end{array}
$$

We also see that:

$$234_5$$
$$+ \quad 123_5$$
$$\overline{412_5}$$

Example 8.3.3. *Use base 2 addition to find the sum of* 10010100_2 *and* 111011_2.

Solution:

$$10010100_2$$
$$+ \quad 111011_2$$
$$\overline{11001111_2}$$

Problem 8.3.1. *Use base 2 addition to find the sum of* 11011_2 *and* 110011_2 *and* 10111_2, *then convert this sum to a base 10 number.*

Problem 8.3.2. *Use base 3 addition to find the sum of* 12210_3 *and* 22112_3.

Problem 8.3.3. *Use base 5 addition to find the sum of the following numbers:* 2334_5 *and* 10124_5 *and* 43210_5. *Then convert this sum to a base 10 number.*

We are now ready to see how to play Nim correctly.

Example 8.3.4. *Suppose we have the starting position* $< 3, 4, 5 >_1$ *for a Nim game. What value do we associate to this game position? How do we decide on the best move at any given time?*

Solution: With each pile we associate as a value, the number of chips in that pile. The value of the game position is determined from the *Nim sum*, denoted $\oplus$, of the values of all the piles. But we do this Nim sum in a special way.

$$3_{10} = 011_2$$
$$4_{10} = 100_2 \text{ and}$$
$$5_{10} = 101_2$$

Suppose we are player 1 in the Nim game $< 3, 4, 5 >_1$. We must decide what to do on our first move. We consider the base 2 (binary) representation for the value of each pile.

Now we do a special Nim addition on these base 2 numbers, which is base 2 addition done without carrying digits to the next column. Essentially, we add the columns *mod* 2. This produces the value 010_2. Note that Nim addition is the same as pairing 1s in a column and canceling all such pairs; that is, if there is an even number of 1s in a column then the column sum is 0 and it is 1 otherwise. In computer science, this is called the *exclusive or* operation and is denoted *xor*.

Our position value according to this Nim sum is 010_2, which is not 0 (this is an **N**-position). Our goal is to put our opponent into a zero position; that is, to eventually leave them with an empty board, the terminal position. Before we can do that, we wish to put them into a losing position, that is, a position from which they cannot avoid the terminal position. We can do that if we can remove chips from one pile and leave the Nim sum of the remaining piles as $00\ldots0_2$ (and hence a **P**-position). In our example, removing 2 chips from pile 1 gives us:

$$
\begin{array}{rcl}
1_{10} & = & 001_2 \\
4_{10} & = & 100_2 \\
\oplus \quad 5_{10} & = & 101_2 \\
\hline
& & 000_2
\end{array}
$$

Thus, we get a zero value for this position (as we did for the empty board). The nice feature of this position is that no matter what move our opponent makes, the resulting board will have nonzero value. This is because some column of one of the binary numbers must change and thus the parity of the number of 1s in that column must change, no longer yielding an even number of 1s. Now, we will move when the board has nonzero value and hence cannot be empty. Our goal will be to force our opponent to always play from a board with Nim sum value zero. Thus, we will always be playing on a board with nonzero value. Eventually, the zero valued board for our opponent will be the empty board and we will win.

Let us continue with the example. Faced with the board above, say our opponent removes all 5 chips from the third pile reducing the game to $< 1, 4, 0 >$. We will respond by removing three chips from the second pile producing the game $< 1, 1, 0 >$, which clearly has value zero. Further, since it is our opponent's move, we know from the Lemma that we shall win.

Our opponent's move above when faced with $< 1, 4, 5 >$ did not matter at all. No matter what their move, we would have been able to return them to the zero position and eventually win the game.   □

We also see that **P**-positions go only to **N**-positions, which in turn always lead to **P**-positions.

Example 8.3.5. *Determine who should win the following game:* $< 2, 3, 4, 5 >$. *Who should win the game* $< 3, 3, 7, 8 >$?

Solution: In the first game we see that the initial value of the game is:

$$
\begin{array}{rcl}
2_{10} & = & 010_2 \\
3_{10} & = & 011_2 \\
4_{10} & = & 100_2 \\
\oplus\ 5_{10} & = & 101_2 \\
\hline
 & & 000_2
\end{array}
$$

Thus, the first player is faced with a zero board and should lose.

In the second game we see the initial board value is clearly nonzero. This is because the two piles of 3 chips will cancel each other out, leaving two distinct values of $7_{10} = 111_2$ and $8_{10} = 1000_2$. Clearly their Nim sum is nonzero. Thus, the first player will be able to win.   □

8.3.1   Poker Nim

An easy variation to Nim is called *Poker Nim*. The rules of the game are exactly the same with one exception. In place of a move requiring you to take some number of chips from one of the piles, you may instead place back on exactly one pile, any number of chips that you have earlier removed from the game. Clearly, your ability to use this option depends upon how many chips you have already removed and how many of these you have left.

Problem 8.3.4. *Find an optimal strategy for Poker Nim.*

8.3.2 Moore's Nim

Another more complex variation of Nim is called *Moore's Nim* after E. H. Moore [28] who suggested the game. The central variation here is that upon your turn you may reduce the size of any positive number up to k of the heaps, for some fixed $k \geq 1$. The amount reduced in each pile can vary. Clearly, $k = 1$ is just ordinary Nim. Moore's Nim will be denoted NIM_k.

The remarkable fact is that the strategy for Nim "generalizes" to NIM_k. The way to proceed is as follows:

- Find the value of each pile in base 2.

- Add the base 2 numbers in base $k + 1$ without carrying digits, that is, the addition is done *mod* $k + 1$, and denoted $\oplus_{k+1}$.

- Move to a position in which this NIM_k sum is zero.

Note that in ordinary Nim we did addition in base $2 = k + 1$ and we did not carry digits. Thus, the strategy is a generalization of the Nim addition rule.

Example 8.3.6. *Suppose we are playing Moore's Nim with $k = 2$ and a board with piles of 12, 13, 14 and 15 chips. What is the proper first play?*

Solution: Following our new strategy, we proceed exactly as in Nim and first find the base 2 representation for the number of chips in each pile. Then, since $k = 2$, our second step is to add these numbers in base 3 (clearly using only the digits 0, 1 and 2). Doing this we obtain:

$$12_{10} = 1100_2$$
$$13_{10} = 1101_2$$
$$14_{10} = 1110_2$$
$$\oplus_3 \quad 15_{10} = 1111_2$$
$$\overline{\qquad\qquad 1122_3}$$

Our third step is to remove chips from one or two piles in such a way as to reduce this sum to zero mod 3. We can do this in several ways. One such way is by removing 9 chips from the pile of 12 chips. If we do this our new sum becomes:

$$3_{10} = 0011_2$$
$$13_{10} = 1101_2$$
$$14_{10} = 1110_2$$
$$\oplus_3 \quad 15_{10} = 1111_2$$
$$\overline{\qquad\qquad 0000_3}$$

Once we have placed our opponent in the NIM_3 zero position, we play as we did in ordinary Nim. Simply continue to place the opponent back in the NIM_3 zero position with each subsequent move. It should be clear that our opponent, faced with the zero sum, must make a move to a nonzero sum as at most k of the values have columns that change.   □

We also note here that in Nim_k, it is $k+1$ identical piles that are equivalent to zero (as we obtained above) rather than just two identical piles, as in ordinary Nim (where $k = 1$).

Example 8.3.7. *We are playing Moore's Nim with $k = 3$. Find the best first move when the board is piles of 5, 6, 9 and 14 chips.*

Solution: We proceed as before in finding the base two expansions of the chip counts.

Now it is easy to see that one possible move is to reduce the piles of 6, 9 and 14 chips to 5 chips each. Then there will be 4 piles of 5 chips each and in base four, our sum must be zero.   □

$$\begin{array}{rl}
5_{10} & = 0101_2 \\
6_{10} & = 0110_2 \\
9_{10} & = 1001_2 \\
\oplus_4 \quad 14_{10} & = 1110_2 \\
\hline
& 2322_4
\end{array}$$

8.3.3 Other Games

Northcott's Game is played on a checkerboard which has one white and one black checker in each row (see Figure 8.1). You may move any one checker in your color to another empty square in that row. However, you may not jump over your opponent's checker in any row. If you cannot move, you lose.

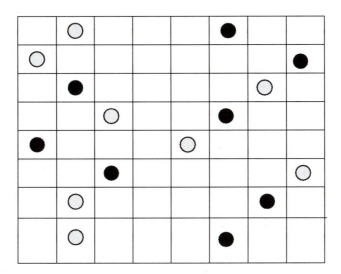

FIGURE 8.1: An example of Northcott's game.

Problem 8.3.5. *Determine a strategy for Northcott's Game.*

The Silver Dollar Game is played on a semi-infinite strip of squares with a finite number of coins, no one of which is a silver dollar (see Figure 8.2). Each coin is placed on a separate square. A legal move is to move a coin to the left (towards the finite end of the strip). During your move you may not jump over or land on any other coin. The game ends as usual, when one player has no move. This happens because the coins are all jammed to the left (finite) end of the strip.

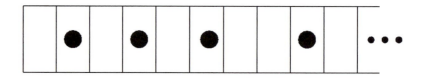

FIGURE 8.2:   The Silver Dollar game.

Problem 8.3.6. *Determine a strategy for the Silver Dollar Game.*

There is a variation of the silver dollar game in which one of the coins is a silver dollar. It appears randomly placed among the coins on the strip. Now the left end square is replaced with a bag and a move could be to place the left most coin in the bag. That coin is now done moving. When the silver dollar is put in the bag, the game ends with the person placing the silver dollar in the bag the loser. The other person collects the bag as the prize for winning!

Problem 8.3.7. *Determine a strategy for this game. (Hint: Think about every other interval between coins, starting for the right most coin.)*

Exercises:

8.3.1 Find the base 2 representation for 33_{10}, 18_{10} and 54_{10}. Find the base 4 and base 5 representations for 67_{10}.

8.3.2 What is a proper first move in the Nim game with piles of 5, 7, 8 and 9 chips.

8.3.3 Which player should win the Nim game with piles of 2, 4 and 6 chips?

8.3.4 Find a proper first move in the Nim game with piles of 2, 4, 6 and 8 chips.

8.3.5 Find the base 5 Nim sum for the following base 2 numbers:
  (1) 110101, 1010101, 1100110, 111000, and 1111111.
  (2) Now find the base 4 Nim sum for the same numbers.
  (3) Now find the base 3 Nim sum for the same numbers.
  (4) Now find the base 2 Nim sum for the same numbers.

8.3.6 Given the Nim piles from the previous problem, what is a proper first move when
  (1) $k = 2$,
  (2) $k = 3$, and
  (3) $k = 4$?

8.3.7 Given chip piles of 11, 12, 13 and 14 chips, determine a best first move for Moore's Nim with $k = 1$, $k = 2$ and $k = 3$.

8.3.8 If Nim piles of size 9_{10} and x_{10} have a Nim sum of 101_2, then what is x?

8.3.9 If Nim piles of sizes 11_{10}, 14_{10} and x_{10} have a Nim sum of 0_2, then what is x?

8.3.10 Eight vertices of a graph are placed along a straight line. Two players alternate drawing an edge between two consecutive vertices. The first player to form a path on three vertices loses. Show that player 1 has a winning strategy for this game. What changes if there are 10 vertices?

8.4 Games as Digraphs

We next wish to model games in yet another way, as directed graphs. This is done by associating a unique vertex of the digraph with each possible position of the game. If there is a move that takes the game from position A to position B, then we insert a directed edge from the vertex corresponding to A to the vertex corresponding to B.

Example 8.4.1. *Consider the subtraction game with $S = \{1, 2, 3\}$ on a pile of 8 chips. What is the corresponding game graph?*

Solution: We have a digraph with 9 vertices, 8 from the pile and 1 representing the empty pile. We label these vertices as 0 to 8. Then, any vertex will have at most three directed edges out to the vertices labeled with the next three smallest values (if they exist). The graph for this game is shown in Figure 8.3.   ☐

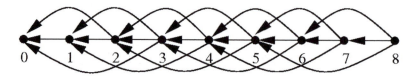

FIGURE 8.3:   The digraph of the subtraction game.

We shall consider digraphs as ordered pairs (V, F) where V is the vertex set of the digraphs and F is a function that gives for each $x \in V$, a subset $F(x) \subset V$, called the *followers* of x which are those vertices directly adjacent from x via a directed edge. Such vertices are sometimes called *out neighbors* or *children* of x. If $F(x)$ is empty, then x is the terminal position.

The **P**- and **N**-positions in a game digraph can be characterized inductively as follows:

- A vertex v is a **P**-position if and only if all its followers are **N**-positions.

- A vertex v is an **N**-position if and only if it has some follower that is a **P**-position.

The induction starts at the *sinks* (that is, vertices with no followers) which are (vacuously) **P**-positions. We shall see much more about this later.

Example 8.4.2. *Consider the subtraction game with $S = \{1, 3\}$ on 7 vertices. Find the game digraph for this game.*

Solution: The game digraph is shown below.

8.4.1   Sums of Games

In this section we define the natural operation of combining games to make bigger games. More formally, if G_1 and G_2 are impartial games, then their sum $G_1 + G_2$ is another impartial game played as follows: on

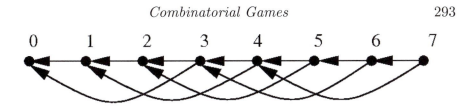

FIGURE 8.4: The digraph when $S = \{1, 3\}$.

each turn one player chooses one of G_1 or G_2 and plays on that game, leaving the other game untouched. The game ends when no moves are possible on either G_1 or G_2.

If $G_1 = (V_1, E)$ and $G_2 = (V_2, F)$ are game digraphs, then their sum $G_{sum} = (V, F^*)$ where $V = V_1 \times V_2$ and the directed edge set is defined as

$$F^* = F(v_1, v_2) = E(v_1) \times \{v_2\} \cup \{v_1\} \times F(v_2)$$

Thus, a move from (v_1, v_2) is really just a move on one of the individual games G_1 or G_2.

Clearly, simple subtraction games become more interesting as Nim type games with several piles. These are just the sum of several games. One might naturally ask if knowing the **P**-positions and **N**-positions for G_1 and G_2 is enough to tell us how to play $G_1 + G_2$ properly. Unfortunately, it is not enough. We need to generalize the notion of these positions, which is known as the Sprague-Grundy function.

8.4.2 The Sprague-Grundy Function

We can analyze game digraphs through P-positions and N-positions as before, but we need more for game sums. We use another device called the *Sprague-Grundy function*, which can generally tell us more. Before defining this function we need another definition.

Let $N = \{0, 1, 2, \ldots\}$ be the set of *natural numbers*. If S is a proper subset of N then let *mex* S be the smallest natural number not in S (the minimum excluded value); that is,

$$mex\ S = min\ (N - S).$$

Definition 8.4.1. *The Sprague-Grundy function of a digraph (V, F) is a function sg defined on V and taking nonnegative integer values such that*

$$sg(x) = mex\{\ sg(y)\ :\ y \in F(x)\}.$$

We should note that $sg(x)$ is recursively defined; that is, $sg(x)$ is defined in terms of $sg(y)$ for followers y of x. For terminal vertices $sg(x) = 0$. For nonterminal vertices x all of whose followers are terminal, $sg(x) = 1$ and so forth.

This process works for graphs that are *progressively bounded*, which means that no matter what vertex we start at, any path to a terminal vertex has length bounded by some number n. We shall limit our attention to game graphs of this type, as game graphs with cycles are more complicated and do not fit our assumption that there can be no ties.

Given the Sprague-Grundy function sg of a digraph, the P-positions correspond to positions with $sg(x) = 0$ and all other positions are N-positions. This is easy to check using the definitions.

Example 8.4.3. *Determine the Sprague-Grundy function for the subtraction game with $S = \{1, 2, 3\}$.*

Solution: Using our example graph of Figure 8.3 we see that vertex 1 can only move to vertex 0 which is the terminal vertex with $sg(0) = 0$. Thus, $sg(1) = 1$. A similar argument implies $sg(2) = 2$ and $sg(3) = 3$. But vertex 4 can only move to $1, 2$ or 3, and so by the mex rule, $sg(4) = 0$. Overall, sg is as shown below.

x	0	1	2	3	4	5	6	7	8
sg(x)	0	1	2	3	0	1	2	3	0

□

The Sprague-Grundy function satisfies two important properties.

Theorem 8.4.1.

- *Vertex v is a **P**-position if and only if $sg(v) = 0$.*

- *If $G = G_1 + G_2$ and $v = (v_1, v_2)$ is a position in G, then $sg(v)$ is the Nim sum of $sg(v_1)$ and $sg(v_2)$, that is,*

$$sg(v) = sg(v_1) \oplus sg(v_2).$$

(*Laskar's Nim*) This Nim variation is due to Emanuel Laskar, world chess champion from 1894 to 1921. The game is as follows. Suppose each player is allowed on any turn to

(1) Remove any number of chips from one pile as usual, or

(2) Split one pile containing at least two chips into two piles (with no chips being removed).

Example 8.4.4. *Find the Sprague-Grundy function for Laskar's Nim.*

Solution: Clearly, the Sprague-Grundy function for the one pile game satisfies $sg(0) = 0$ and $sg(1) = 1$. The followers of 2 are 0 and 1 and $< 1, 1 >$. The Sprague-Grundy values on these are 0, 1 and $1 \oplus 1 = 0$, hence $sg(2) = 2$. Similarly, the followers of 3 are 0, 1, and 2 as well as $< 1, 2 >$. These have values $0, 1, 2$ and $1 \oplus 2 = 3$ and so $sg(3) = 4$. Continuing in this manner we find that:

x	0	1	2	3	4	5	6	7	8	9	10	...
$sg(x)$	0	1	2	4	3	6	5	8	7	10	9	...

□

The Game of Kayles: This game was introduced by Sam Lloyd and H. E. Dudney.

Two bowlers face a line of $n > 2$ bowling pins in a row. The players are good enough that they can knock down any single kayle pin, or any two adjacent kayle pins. But pins are spaced so that no player can do better than knock down two consecutive pins. The first player facing no pins to knock down loses.

An alternate description for this game would be the game is played with piles of chips and the allowed moves are such that you may remove one or two chips from the pile and then, if you wish, you may split the pile into two piles. These correspond to knocking down one or two pins and splitting or not splitting the pile means the pins you knocked down were in the middle of the row or at the end of the row.

Example 8.4.5. *Find the Sprague-Grundy function for the game of Kayles.*

Solution: The only terminal position is a row of no pins. Thus $sg(0) = 0$. One pin can only be reduced to the empty row so $sg(1) = 1$. A row of two pins can be reduced to either a row of one pin or no pins, thus $sg(2) = 2$. A row of three pins can be reduced to rows of 2, 1 or to two rows of 1 pin each. Thus, $sg(3) = 3$.

The early values the Sprague-Grundy function for Kayles are shown below. The values are periodic from $n = 72$ on, repeating every 12th value.

$$
\begin{array}{c|ccccccccccc}
x & 0 & 1 & 2 & 3 & 4 & 5 & 6 & 7 & 8 & 9 & 10 \\
sg(x) & 0 & 1 & 2 & 3 & 1 & 4 & 3 & 2 & 1 & 4 & 2
\end{array}
$$

8.4.3   More about Impartial Games

Recall that an impartial game is one in which the options for the two players are always the same as sets. In this section, we state two important results about impartial games. See [12] for proofs of these results. These two results show the central position of Nim in such games.

Theorem 8.4.2. *Let G be a game played with a finite collection of numbers (from 0, 1, 2, ...) in the following way. Each move affects just one number and strictly changes that number. Any decrease of a number is always attainable by a legal move, but some increases may also be possible. However, the rules of the game are such as to ensure that the game terminates. Then any outcome of any position in G is the same as that of the corresponding position in Nim.*

A *short game* is one that has only finitely many positions possible.

Theorem 8.4.3. (*Grundy's Theorem*) *Each short impartial game G is equivalent in play to some Nim heap.*

Exercises:

8.4.1 Find the Sprague-Grundy function for the graphs below.

8.4.2 Suppose we play the subtraction game with the rule you must remove at least half the pile. What is the Sprague-Grundy function if the pile has 11 chips?

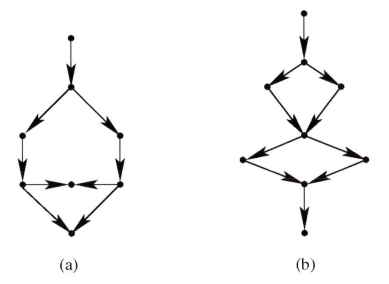

(a)   (b)

FIGURE 8.5: Two Sprague-Grundy graphs.

8.4.3 What is the Sprague-Grundy function for the subtraction game with $S = \{1, 3, 5\}$ and the pile has 21 chips?

8.4.4 Suppose in the previous problem we change S to be $S = \{2, 4, 6\}$. Now what is the Sprague-Grundy function?

8.4.5 Find the Sprague-Grundy function of the subtraction game with subtraction set $S = \{1, 3, 4\}$.

8.4.6 Consider the subtraction game where you may take away any even number of chips, or one chip if the pile is just one chip. Find the Sprague-Grundy function for this game.

8.4.7 (*Wythoff's Game*) This game is played on a standard 8×8 chessboard. Players take turns moving one queen. But the queen may only move vertically down, or horizontally to the left or diagonally down. When the queen reaches the lower left corner, the game is over. Find the Sprague-Grundy function by writing the value in each square of the chessboard.

8.4.8 Suppose Wythoff's game is played with a rook instead of a queen. The rook moves only vertically down or horizontally to the left. Now find the Sprague-Grundy function.

8.4.9 Suppose you are playing Kayles with 7 pins. What is the correct first move?

8.4.10 Suppose you are playing Kayles with 13 pins and the second pin is already knocked down.
(1) Show this is an **N**-position.
(2) Find a winning move (that is, which pins you should knock down).

8.4.11* Suppose you are playing the subtraction game with $S = \{1, 2, 3\}$. If you are playing the sum of three games with piles of 4, 5 and 6 chips, find the Sprague-Grundy function for this game.

8.4.12 Consider a subtraction game where the rules are, from any one pile you may remove any even number of chips, provided it is not the entire pile, or the whole pile provided the pile is odd in number.
(1) Determine the Sprague-Grundy function for the one pile game.
(2)* Suppose the game is played with three piles of 5, 10 and 15 chips. Now find the Sprague-Grundy function for this game.

8.5   Blue-Red Hackenbush

In this section we will consider another combinatorial game called *Blue-Red Hackenbush*. The board for this game is a line drawing using finite blue and red line segments. Any drawing is allowed provided there is a connection to the ground (see Figure 8.6). Players alternately remove an edge of their color (one player is blue and one player is red). An edge of any color is also removed provided it has become disconnected from the ground. As usual, the first player with no move loses.

We next wish to assign values to these games, in a manner somewhat similar to what we did in Nim. Each Nim pile had a value and the game value was the Nim sum of the piles. Here, each separate drawing will have a value and we can combine values to determine the overall board value. A positive value will indicate a blue advantage, while a negative value indicates a red advantage. But what we shall see is that the values in Hackenbush can be much more general than those of Nim.

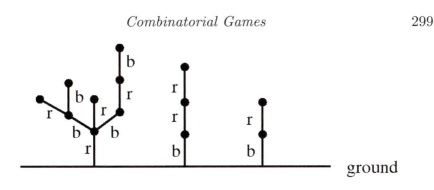

FIGURE 8.6:   A blue-red Hackenbush game.

Let's begin with some simple examples and develop a feel for the values we will assign.

Example 8.5.1. *Given a blue-red Hackenbush game shown in Figure 8.7, which player will win? That is, is there an advantage for one player over the other?*

FIGURE 8.7:   A blue-red Hackenbush example.

Solution: If the blue player moves first, the only blue edge is removed and the only red edge is also removed as it is disconnected from the ground. Thus, blue will win.

If red goes first, red removes the only red edge. Blue then removes the only blue edge and red is faced with the empty board and loses. So again blue has won. Thus, there is a clear advantage for the blue player in this game.   □

But how big is this advantage? Could this advantage be 1? We can test this hypothesis as follows. If blue has an advantage of 1 in this game, then adding one independent red edge (see Figure 8.8) should balance this advantage.

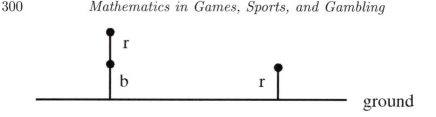

FIGURE 8.8:   A test of the one edge advantage hypothesis.

Now who wins? If red starts, then red will remove the upper red edge as it is in danger from blue. Then blue will remove its one edge and finally red removes the independent red edge. Now blue faces the empty board and loses. In this case red has the advantage.

Next suppose that blue starts. Then blue removes the only blue edge and the upper red edge. Red removes the independent red edge. Again blue faces the empty board and loses. Thus, in either case red now has the advantage. We must conclude that the blue advantage in the first example is real, but not as much as 1. Hence, it appears Hackenbush games can have fractional values!

But we still seek the value of the original game from Example 8.5.1. Let's try one more test. Could the blue advantage be 1/2? To test this hypothesis, consider the example below.

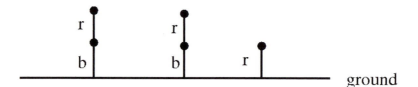

FIGURE 8.9:   A test for the value 1/2.

If the original advantage to blue was 1/2, the two copies of that game and one copy of an independent red edge together should balance to zero, that is no advantage to either color.

To test this hypothesis consider the following play. If blue goes first, then one blue and also one red edge are removed. Now red removes the remaining upper red edge. Blue removes its last edge and red removes the independent red edge. Blue faces the empty board and loses.

If red goes first, red removes one of the upper red edges, say the left most. Blue now removes the right most blue edge, and also the red edge attached above it. Red now removes the red independent edge and blue follows removing the left most blue edge. Red faces the empty board and loses.

Thus, either red or blue can win, depending only on who goes first. In fact, the second player won in each case. We call such a game a *fair game* with game value 0.

We now note that reversing the labels on a Hackenbush game should reverse the value, that is, convert a red advantage to an equal blue advantage or convert a blue advantage to an equal red advantage.

Problem 8.5.1. *Devise a test to prove the value of the Hackenbush game of Figure 8.10 is* 1.5.

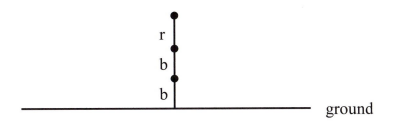

FIGURE 8.10: An advantage of 1.5.

As we think about the values a Hackenbush game can have, we clearly need better methods to determine these values. Guessing at values and testing as we have done will prove increasingly difficult. We need an algorithm to convert *Hackenbush strings*, that is the simple drawings that stack one edge after another, to their number value. To accomplish this we make the following conventions. Positive values are blue (or Left) advantages, while negative values are red (or Right) advantages.

To accomplish our goal, we need one more extension to our number bases, namely to expand fractional values in base two. This turns out to be a natural extension of the base form for integers. Recall a base 2 representation for an integer m is

$$m = a_k \times 2^k + a_{k-1} \times 2^{k-1} + \ldots + a_1 \times 2 + a_0.$$

For fractional values, why not just extend the powers of 2 to negative values? That is, we can also consider

$$b_1 \times 2^{-1} + b_2 \times 2^{-2} + \ldots + b_k \times 2^{-k}.$$

Example 8.5.2. *Find a base 2 representation for the fraction $\frac{5}{8}$.*

Solution: Since $\frac{5}{8} < 1$, the integer part of our number is clearly zero. Now we simply need to find $\frac{5}{8}$ as the sum of these powers of 2. We note that

$$\frac{5}{8} = \frac{1}{2} + \frac{1}{8}.$$

Thus, a base 2 representation for $\frac{5}{8}$ is

$$1 \times \frac{1}{2} + 0 \times \frac{1}{4} + 1 \times \frac{1}{8}$$

so that

$$\left(\frac{5}{8}\right)_{10} = 0.101_2.$$

$\square$

Example 8.5.3. *Find the binary representation for $\left(2\frac{3}{4}\right)_{10}$.*

Solution: To find the representation we consider the integer and fractional parts separately, then we put the two pieces together. Clearly $2_{10} = 10_2$. Now for the fractional part $\frac{3}{4}$, we note that

$$\left(\frac{3}{4}\right)_{10} = \left(\frac{1}{2}\right)_{10} + \left(\frac{1}{4}\right)_{10} = 1 \times \frac{1}{2} + 1 \times \frac{1}{4}.$$

Thus,

$$\left(\frac{3}{4}\right)_{10} = 0.11_2.$$

Putting the two pieces together we see that the base 2 representation is

$$\left(2\frac{3}{4}\right)_{10} = 10.11_2.$$

$\square$

The following Hackenbush conversion algorithm is due to E. R. Berlekamp [5].

Example 8.5.4. *Convert the number $2\frac{3}{4}$ to an equivalent valued blue-red Hackenbush string.*

Solution: The first step in the algorithm is to convert the integer part of the number to a Hackenbush string. As 2 is positive, the conversion is to 2 blue edges, the first attached to the ground (see Figure 8.11).

Next we make the conversion of the fractional part. To indicate the part of the string that represents the fraction, we attach a blue and then a red edge to the end of the string we have already built. This blue edge to first red edge is a flag to indicate the binary point, not a decimal point as we are working in base 2.

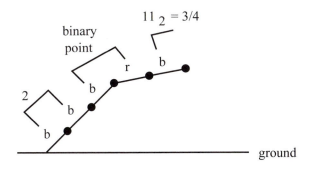

FIGURE 8.11: The Hackenbush string for $2\frac{3}{4}$.

Now we attach the fractional binary expansion onto the string, using a blue edge to indicate a 1 in the string and a red edge to indicate a 0 in the string, omitting the final 1 of the string. In our case $\frac{3}{4}_{10} = 11_2$ so, ignoring the final 1 we attach one blue edge to the string (see Figure 8.11). □

We now note that if the number had been negative, we would follow the same algorithm, except reversing the roles of red and blue. Thus, we must always keep in mind whether the number is positive or negative.

Example 8.5.5. *Find the Hackenbush string with value* $-1\frac{9}{16}$.

Solution: This time our number is negative so we reverse the color roles. The integer part converts to one red edge. The binary point is then indicated with one red edge followed by one blue edge (this is the convention when the number is negative). Finally, the fractional part has a base 2 representation of $.1001_2$. The Hackenbush string thus obtained is shown in Figure 8.12. □

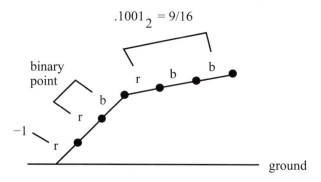

FIGURE 8.12:   Hackenbush string for $-\frac{1}{4}$.

These Hackenbush strings can be combined to form blue-red Hackenbush games. The value of such a game is just the sum of the values of the independent strings.

Example 8.5.6. *Find the value of the Hackenbush game shown in Figure 8.13.*

Solution: We apply the Berlekamp rule to each string. The left most string converts to $1\frac{1}{2}$ and the rightmost string converts to $-1\frac{3}{4}$ so the value of the game is $-\frac{1}{4}$.   □

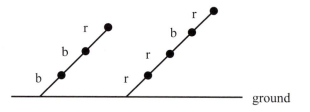

FIGURE 8.13:   Hackenbush string for $-1\frac{9}{16}$.

Exercises:

8.5.1 Convert each of the following numbers to a red-blue Hackenbush string.

(1) 4.25   (2) −3.125   (3) $\frac{1}{16}$

(4) $-\frac{3}{16}$ (5) 0.125.   (6) 2.5

8.5.2 Apply the Berlekamp algorithm to find the value of the following red-blue Hackenbush strings.

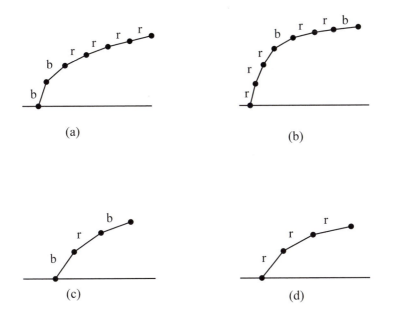

FIGURE 8.14:   Four examples of Hackenbush strings.

8.5.3 Given the game below, find a game to add to it so that the resulting game has value zero.

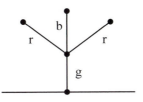

FIGURE 8.15:   Generalized Hackenbush game.

8.6   Green Hackenbush

A simple variation of blue-red Hackenbush is known as *Green Hackenbush*. The variation is simple in that some edges (possibly all) are now colored green and can be chosen by either player at any time. These green edges can create different interesting situations. A very simple example is a green Hackenbush game in which there is only one green edge. It is clear that the first player will remove this edge and win. This is an example of a combinatorial game where the first player wins.

8.6.1   Pruning Green Hackenbush Trees

We have seen that green Hackenbush strings is nothing but Nim in a very poor disguise. But we originally said that Hackenbush could be played on any kind of diagram. In this section we will extend the type of green Hackenbush diagram we can handle.

Suppose we have a green Hackenbush tree; that is, a rooted (to the ground) graph without cycles and all its edges are green (see Figure 8.16).

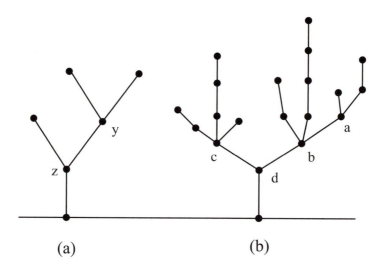

FIGURE 8.16:   Green Hackenbush trees.

As before, a move consists of removing some edge and any other edges that become disconnected from the ground. Since this game is impartial and short, we know by Theorem 8.4.3 that any such tree is equivalent to a Nim pile. But which Nim pile? What we seek is the Sprague-Grundy value of any green tree.

To accomplish this, we use the following principle:

Replacement Principle: When branches of a green Hackenbush tree meet at a vertex, one may replace these branches with one string equal in length to their Nim sum.

This principle amounts to viewing the branches at a vertex as individual Nim strings that amount to a game in themselves. An example or two will help reinforce this idea.

Example 8.6.1. *Find a Nim pile that is equivalent to the green Hackenbush tree of Figure 8.16 (a).*

Solution: There are two branches at vertex y, each with one edge. Since $1 \oplus 1 = 0$, we may simply delete these two branches, or equivalently, replace them with the empty branch.

Now at vertex z, there are also two branches, each of length one. Thus, these branches may be deleted, leaving a single green edge whose value is one. Thus an equivalent Nim pile has one chip. $\qquad\square$

The previous example is quite easy, so let's consider a more involved situation.

Example 8.6.2. *Find an equivalent Nim pile for the green Hackenbush tree of Figure 8.16 (b).*

Solution: The right most branch meeting at vertex a has two branches, one of length one and one of length two. Since $1 \oplus 2 = 3$, we replace those branches with one of length three (see Figure 8.17).

Next we consider the branch rooted at b with branches of lengths two, four and four. Since $2 \oplus 4 \oplus 4 = 2$, we may delete the two branches of length four (see Figure 8.18).

The third step is to reduce the branches rooted at vertex c. Here they have lengths two, three and one. Since $2 \oplus 3 \oplus 1 = 0$, we simply delete these branches.

Finally, we reduce at the branch rooted at vertex d. The new branches rooted at d have lengths one and three. Since $1 \oplus 3 = 2$, we replace

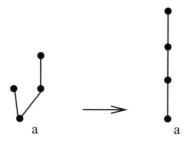

FIGURE 8.17:   Step 1 in finding the equivalent Nim pile.

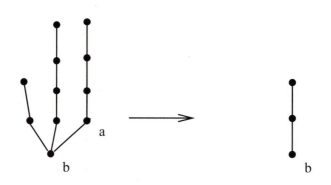

FIGURE 8.18:   Step 2 in finding the equivalent Nim pile.

these branches with one branch of length two. Hence, the equivalent Nim pile has exactly three chips.   □

   Although there are techniques for handling even more complex green Hackenbush games, we will refer the reader to [7] for more on the subject.

Exercises:

8.6.1 Apply the Reduction Principle to find the equilivent Nim pile for each of the green Hackenbush trees in Figure 8.19.

8.6.2 You are playing green Hackenbush and the initial board consists of the two trees shown in Figure 8.20. Find the best opening move for this game.

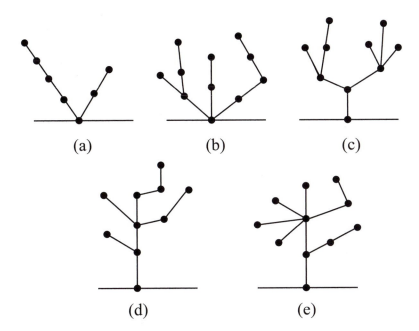

FIGURE 8.19:   Five Hackenbush trees.

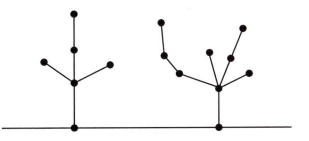

FIGURE 8.20:   A Hackenbush game.

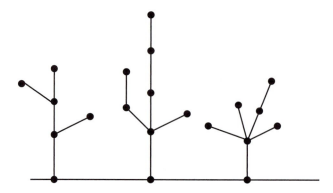

FIGURE 8.21:   Another Hackenbush game.

8.6.3 If confronted with the green Hackenbush game of Figure 8.21, find the best opening move.

8.6.4 Create a green Hackenbush tree game that is a first player wins game.

8.6.5 Create a green Hackenbush game that is a second player wins game.

8.7   Games as Numbers

In this section we will take a more theoretical look at combinatorial games. This is the little bit of theory promised earlier.

A *game* is an ordered pair of sets of games. We usually write the game G as

$$G = \{\, \{G^{L_1}, G^{L_2}, \ldots\} \mid \{G^{R_1}, G^{R_2}, \ldots\} \,\}$$

or more simply as

$$G = \{\, G^L \mid G^R \,\}$$

where G^L and G^R are sets of games that will occur (possibly empty) due to a move from the Left (blue) or Right (red) player.

The *end game* is

$$\{\, \emptyset \mid \emptyset \,\} = \{\, \mid \,\}$$

in which neither player has any moves. We denote this game 0, thus,

$$0 = \{\, \emptyset \mid \emptyset \,\}.$$

The Hackenbush game with exactly one blue edge produces

$$\{\, 0 \mid \,\}$$

as any move by Blue (Left) produces an empty board and hence the zero game, while Red (Right) has no move. This game has value 1 and hence,

$$1 = \{\, 0 \mid \,\}.$$

Similarly,

$$-1 = \{\, \mid 0 \,\}.$$

We now define game

$$* = \{\, 0 \mid 0 \,\}.$$

Here, each player has a move that takes us to the zero game. For example, the Hackenbush game with one green edge is such a game. Note that this is a first player wins game.

We have now seen the four fundamental game types. All other games will be comparable to one of these. The four types are shown in Table 8.1. Every game belongs to one of these four classes.

TABLE 8.1:   Four Possible Game Types

ZERO GAMES	NEGATIVE GAMES
$G = 0$	$G < 0$
2nd player wins	Right (or red) wins
POSITIVE GAMES	FUZZY GAMES
$G > 0$	$G \parallel 0$
Left (or blue) wins	First player wins

Other values are fairly easy to see. We consider some of these now. For example,

$$\frac{1}{2} = \{\, 0 \mid 1 \,\}.$$

The blue-red Hackenbush game with one red edge on top of one blue edge was such a game. Blue's only move creates the empty board, that is, the zero game, while red's only move leaves one blue edge, hence the game 1.

Now it is also easy to see that

$$1\frac{1}{2} = \{\, 1 \mid 2 \,\}.$$

We use the Hackenbush game with one red edge above two blue's edges as an example. Red's only move leaves a board with two blue edges, hence the game 2, while Blue's best move is to remove the middle blue edge, leaving a game with one blue edge, hence leaving the game 1.

We note that in general

$$\{\, n \mid n+1 \,\} = n + \frac{1}{2}$$

and

$$\{\, n \mid \,\} = n + 1.$$

Further, the negative of any game (number) can be obtained by reversing the left and right game options. Thus, for example,

$$\{\, \mid -n \,\} = -(n+1).$$

In general, given game $G = \{\, G^L \mid G^R \,\}$, then

$$-G = \{\, -G^R \mid -G^L \,\}.$$

As we know, games can be added. Given $G = \{\, G^L \mid G^R \,\}$ and $H = \{\, H^L \mid H^R \,\}$ we define

$$G + H = \{\, G^L + H, G + H^L \mid G^R + H, G + H^R \,\}.$$

As an example, suppose $G = \frac{1}{2} = \{\, 0 \mid 1 \,\}$ and $H = 1\frac{1}{2} = \{\, 1 \mid 2 \,\}$, then what is $G + H$? By the definition above,

$$G^L + H = 0 + 1\frac{1}{2} = 1\frac{1}{2}$$

and

$$G + H^L = \frac{1}{2} + 1 = 1\frac{1}{2},$$

while

$$G^R + H = 1 + 1\frac{1}{2} = 2\frac{1}{2}$$

and

$$G + H^R = \frac{1}{2} + 2 = 2\frac{1}{2}.$$

Therefore,

$$G + H = \{\, 1\frac{1}{2} \mid 2\frac{1}{2} \,\}.$$

TABLE 8.2: Four Possible Game Types Re-Examined

		If Left Starts	
		Left Wins	Right Wins
If		POSITIVE GAMES	ZERO GAMES
Right	Left Wins	Left wins	2nd player wins
Starts		FUZZY GAMES	NEGATIVE GAMES
	Right Wins	1st player wins	Right wins

Let's reexamine our table of four classes of games in light of who plays first. Again Left (Blue) and Right (Red) are the players. To summarize what is in the table, we describe these four cases based on the player with the winning strategy. We have that

- $G > 0$ or G is positive if Left can always win.

- $G < 0$ or G is negative if Right can always win.

- $G = 0$ or G is zero if the second player can always win.

- $G \parallel 0$ or G is fuzzy if the first player can always win.

As a note, the game of green Hackenbush with one string of green edges is an example of a Fuzzy game. The first player to move removes the green edge and the second player loses. There are clearly infinitely many such games. But the fuzzy game in Figure 8.22 (b) is the sum of

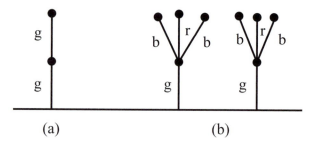

FIGURE 8.22:   (a) A fuzzy game and (b) the sum of two fuzzy games.

two fuzzy games. We note that this game has a positive value. There are enough blue edges that Blue can force Red to take the first green edge. Blue then takes the second green edge and wins.

Thus, addition of games presents some interesting twists. Even more unusual things are possible. To see this, recall that $* = \{ 0 \mid 0 \}$. This is a game, but $*$ is not a number. However, we can certainly add $*$ to any game. In fact, following the definitions we see that

$$x + * = \{ x \mid x \}$$

for any number x. This type of value occurs so often that we give it the notation $x*$, that is

$$x* = x + *.$$

Note that this is following ordinary conventions, as you would write $4\frac{1}{2}$ instead of $4 + \frac{1}{2}$.

Suppose we again consider green Hackenbush. Suppose we have a green Hackenbush game with only green chains. This game is clearly impartial as the same moves are available to each player. The chains are just like Nim piles. We will write $*n$ as the value of any green chain with n edges. Note, this is exactly like a Nim pile with n chips. Then, using the definitions, we see that

$$\begin{aligned}
*0 &= \{ \mid \} = 0, \\
*1 &= \{ *0 \mid *0 \} = \{ 0 \mid 0 \}, \\
*2 &= \{ *0, *1 \mid *0, *1 \} = \{ 0, * \mid 0, * \}, \ldots \\
*n &= \{ *0, *1, *2, \ldots, *(n-1) \mid *0, *1, \ldots, *(n-1) \}.
\end{aligned}$$

These special values are called *nimbers* and they are important to the theory of games. Note that the same options appear on both sides, so these are values for impartial games.

Exercises:

8.7.1 Suppose that $G = 1.5$ and $H = 2$, find $G + H$.

8.7.2 Suppose that $G = -1.5$ and $H = 2.5$, find $G + H$.

8.7.3 Given the games $G_1 = \{*0, *1 \mid *0, *1\}$ and $G_2 = \{*0, *1, *2 \mid *0\}$, what is the game $G_1 + G_2$?

8.7.4 Show that for all games G, that $G + 0 = G$.

8.7.5 Show that for all games G and H that $G + H = H + G$.

8.7.6 Show, using the definitions, that $-(-G) = G$.

8.7.7 Show, using the definitions of this section, that $*0 + *1 + *2 = *3$.

8.8   More about Nimbers

In this section we consider adding nimbers and using nimbers as values for games. When we considered Nim, we assigned the piles a number value based solely on how many chips were in the pile. But we never really assigned the game itself a value. The integer value on a Nim pile was used to determine a game strategy, not fit in with the values of all other games.

When you consider a Nim pile it has a fuzzy value, since the first player can remove the entire pile. We also know two equal piles cancel to a zero value. The sum of two unequal piles is fuzzy, as the first player may make them equal. These remarks imply the following.

Proposition 8.8.1. *In a three-pile Nim game, the first player who equalizes two heaps or removes a heap is the loser.*

But now, suppose we want to find the values for such games. Clearly we must add the individual values for the piles. Suppose we want to add $*5 + *4$. What is the sum? In order to determine that, we must take advantage of what we know about Nim!

Recall that $*x + *x = 0$ because two equal Nim piles is a second player win situation and thus has value 0. We handle $*5 + *4$ by breaking down 5 and 4 in binary form. We know that

$$5 = 4 + 1 = 2^2 + 2^0 \text{ and}$$

$$4 = 4 + 0 = 2^2 + 0.$$

Now each sum has a 2^2 term and we cancel these terms (like equal Nim piles), replacing them with 0. This leaves us with $2^0 + 0 = 1$ as the sum of the remaining terms. Thus,

$$*5 + *4 = *1.$$

Example 8.8.1. *Find the sum* $*12 + *7 + *8$.

Solution: We proceed as before with binary expansions.

$$
\begin{aligned}
12 &= 2^3 &&+2^2 \\
7 &= && 2^2 + 2^1 + 2^0 \\
8 &= 2^3
\end{aligned}
$$

Then we see there are two 2^3 terms which cancel, and two 2^2 terms which cancel. We are left with the sum $2^1 + 2^0 = 3$. Thus,

$$*12 + *7 + *8 = *3.$$

□

Example 8.8.2. *Find the sum* $*13 + *14 + *15$.

Solution: Our process is the same. Find the binary expansions for each term and cancel.

$13 = 2^3 + 2^2 + 2^0$
$14 = 2^3 + 2^2 + 2^1$
$15 = 2^3 + 2^2 + 2^1 + 2^0$

After canceling pairs of like terms we are left with $2^3 + 2^2 = 12$. Thus,

$$*13 + *14 + *15 = *12.$$

□

Thus, we see the addition of nimbers is a straightforward process. In fact, hopefully you noticed the canceling of like pairs of powers of 2 really is nothing more than the Nim sum as done earlier. That is, $5 \oplus 4 = 1_2$ and so $*5 + *4 = *1$.

Exercises:

8.8.1 Find $*3 + *5 + *7$.

8.8.2 Find $*2 + *4 + *11 + *23$.

8.8.3 If $*5 + *x = *3$, then what is x?

8.8.4 If $*11 + *13 + *x = *12$, then what is x?

8.8.5 What is $*6 + *6$?

8.8.6 If $*3 + *x = *2$, then what is x?

Chapter 9

Appendix

9.1 Review of Elementary Set Theory

Our notion of probability will be derived entirely from counting the size of sets and the various operations that one can perform on sets. For us, a *set* will be a collection of objects called *elements*. Sets can be described as a simple listing of each member of the set or by describing the set in more general terms. The set with no elements is called the *empty set* and is denoted by the symbol ∅.

The following are examples of sets:

$$\{1, 2, 3, 4, 5\}$$

$$\{\text{red, white, blue}\}$$

a deck of playing cards without jokers

$$\{1, 4, 9, 16, 25, 36...\}$$

All of the above are perfectly good examples of sets. These are each very different from one another. The first three sets are *finite sets*, which means they contain a fixed number of elements. The fourth set is an *infinite set*. A set is infinite if for any integer r you select, the set contains more than r elements. The third set is defined by description, but it is perfectly clear what its members are, the 52 cards of a standard deck. The last set could also have been described as the squares of all positive integers. The description of a set is by no means unique.

If A and U are sets and every element of U is an element of A, then U is a *subset* of A. A subset X of a set Y is a *proper subset* if Y contains elements not contained in X.

The *sample space* is the universal set in which a given problem is posed. It contains all possible elements for the given situation. The sample space is generally denoted S.

Let $S = \{1, 2, 3, 4, 5\}$ and $U = \{1, 3, 4\}$. Then U is a subset of S. We denote this by $U \subseteq S$. The number 3 is an element of S. We denote this by $3 \in S$. We can negate any shorthand by a slash through the symbol. The number 8 is not a member of S, which is abbreviated by $8 \notin S$.

Now that we have the idea of sets, we would like to be able to manipulate them in useful ways. To that end, we have the following set operations.

For any sets A and B in a sample space S:

- We let $A \cup B$, denote A *union* B, which is the set of all elements in S that are in the set A or in the set B or in both sets (see Figure 9.1).

- We let $A \cap B$, denote A *intersection* B, which is the set of all elements of S that are in the set A and also in the set B.

- We let $\overline{A}$, denote the set A **complement**, which is the set of all elements of S that are not in the set A.

Example 9.1.1. *Consider* $S = \{1, 2, 3, 4, 5, 6, 7, 8, 9, 10\}$, $P = \{1, 7, 8, 9, 10\}$ *and* $Q = \{1, 2, 3, 4, 5\}$. *Find* $P \cap Q$, $P \cup Q$, $\overline{P}$ *and* $\overline{Q}$.

Solution: We have $P \cap Q = \{1\}$ because 1 is the only element which both sets contain.

Also,

$$P \cup Q = \{1, 2, 3, 4, 5, 7, 8, 9, 10\},$$

which is just the set of the elements contained in one set or the other.

The complement of P (or Q) is the set of all elements in the set S, but not in the set P (or Q), hence we obtain

$$\overline{P} = \{2, 3, 4, 5, 6\},$$

$$\overline{Q} = \{6, 7, 8, 9, 10\}.$$

□

Naturally, we're not limited to one operation at a time and we can mix and match to our heart's content. We could also involve more than two sets.

Example 9.1.2. *Now suppose* $D = \{2, 4, 6\}$. *Find* $(P \cap Q) \cup \overline{D}$, $\overline{(P \cup Q)}$, $\overline{(P \cap Q)}$, $(P \cap \overline{Q}) \cup \overline{P}$, *and* $\overline{\overline{P}}$.

Solution: We see that $P \cap Q = \{1\}$, and that $\overline{D} = \{1, 3, 5, 7, 8, 9, 10\}$. Hence, $(P \cap Q) \cup \overline{D} = \{1, 3, 5, 7, 8, 9, 10\}$.

Next,

$$\overline{(P \cup Q)} = \overline{\{1, 7, 8, 9, 10\} \cup \{1, 2, 3, 4, 5\}} = \{6\}.$$

We saw earlier that $P \cap Q = \{1\}$. Hence,

$$\overline{(P \cap Q)} = \{2, 3, 4, 5, 6, 7, 8, 9, 10\}.$$

To find $(P \cap \overline{Q}) \cup \overline{P}$ we note that $P \cap \overline{Q}$ is

$$\{1, 7, 8, 9, 10\} \cap \{6, 7, 8, 9, 10\} = \{7, 8, 9, 10\}.$$

Thus, $(P \cap \overline{Q}) \cup \overline{P}$ is

$$\{7, 8, 9, 10\} \cup \overline{P} = \{7, 8, 9, 10\} \cup \{2, 3, 4, 5, 6\} = \{2, 3, 4, 5, 6, 7, 8, 9, 10\}.$$

Finally, from the definition of complement we see that

$$\overline{\overline{P}} = P.$$

$\square$

It may seem most natural to perform set operations on sets of numbers, but the true power of sets comes when we have sets with more descriptive meanings. For us this will sometimes be in games, sometimes in sports, sometimes in some other setting.

For example, consider the sample space S to be a standard deck of playing cards. Let A be the set of aces, H be the set of cards that are hearts and C be the set of cards that are clubs.

The set $A \cap H$ is the ace of hearts and the set

$$A \cap (H \cup C) = \{\text{ace of hearts, ace of clubs}\}.$$

Sets H and C have no elements in common, thus $H \cap C = \emptyset$.

It may also be very useful when working with set operations to employ Venn diagrams. Venn diagrams can make some set equalities obvious, when the formulas themselves may not be as clear. To help see this, we consider each of the following sets: $A \cup B$ (see Figure 9.1), $B \cap \overline{A}$ (see Figure 9.2) and $\overline{A} \cap \overline{B}$ (see Figure 9.3).

The *order* of a set A is the number of elements in the set, denoted $|A|$. Let S be the set of all positive integers that are even and less than 15. Clearly, $|S| = 7$. If C is the set of playing cards with no jokers,

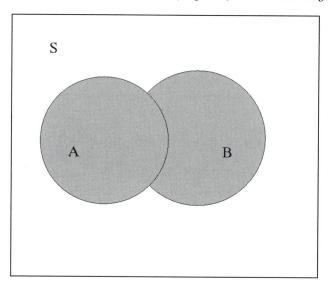

FIGURE 9.1:   Venn diagram for $A \cup B$.

then $|C| = 52$. There is an interesting relationship between the orders of $A \cup B$ and $A \cap B$. When counting the number of elements in A and B, you certainly count each element in either A or B. An element that is in both sets A and B will be counted twice with regard to $|A|$ and $|B|$ and we only wish to count it once with regard to $|A \cup B|$. Thus we get the following formula for any two sets A and B:

$$|A \cup B| = |A| + |B| - |A \cap B|.$$

Example 9.1.3. *On a test* 12 *students made an "A" or answered the bonus question correctly. If* 7 *students made an "A" while* 10 *students answered the bonus correctly, how many students made an "A" and answered the bonus correctly?*

Solution: Using the above result, let A be the set of students who make an "A" and B be the set of students who answered the bonus correctly. Then, the answer to the question will be the order of $A \cap B$.

Applying the equation earns us $12 = 7 + 10 - |A \cap B|$. Accordingly our answer is 5.   □

Example 9.1.4. *Of* 200 *moviegoers surveyed* 123 *always bought candy or popcorn. Of those people,* 45 *always bought only candy while* 51

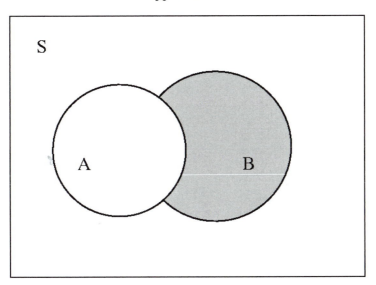

FIGURE 9.2: Venn diagram for $B \cap \overline{A}$.

always bought only popcorn. How many people always buy both popcorn and candy?

Solution: Using Venn diagrams it is easy to compute the intersection to be:

$$123 - (45 + 51) = 27.$$

$\square$

Exercises:

9.1.1 Let $S = \{1, 2, \dots, 15\}$ be the sample space. Let $P = \{2, 4, 6, \dots, 14\}$ and let $Q = \{3, 6, 9, 12, 15\}$. Write out the following subsets:

1. $P \cup Q$.
2. $P \cap Q$.
3. $Q \cap \overline{P}$.
4. $P \cap \overline{Q}$
5. $\overline{P} \cup S$.
6. $P \cup S$.
7. $Q \cap S$.

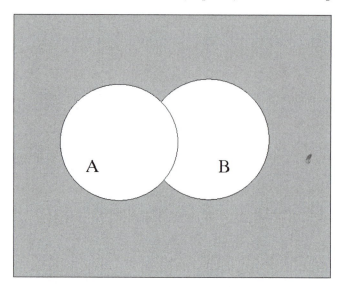

FIGURE 9.3: Venn diagram for $\overline{A} \cap \overline{B}$.

    8. $\overline{P} \cap S$.

9.1.2 Illustrate the following by shading in a Venn diagram:

    1. $\overline{A} \cap B$.

    2. $\overline{A} \cap \overline{B}$.

    3. $\overline{A} \cup \overline{B}$.

    4. $(A \cup B) \cap \overline{C}$.

    5. $(\overline{A \cap B \cap C})$.

What is another way of describing this last set?

9.1.3 Let A be the set of odd positive integers less than 100. Let B be the set of positive integers less than 100 which are divisible by 3. Compute:

    1. $|A \cap B|$.

    2. $|A \cup B|$.

9.1.4 Let H be the set of students who take History 102 and B the set of students who take Biology 112. Assume that $|H| = 40$ and $|B| = 45$. What is $|H \cup B|$ if

1. History 102 and Biology 112 meet at the same time?

2. History 102 and Biology 112 do not meet at the same time and there are 15 students who take both?

9.1.5 Let S be a sample space and A and B be any two subsets of S. Find another way to express each of the following sets.

1. $\overline{A} \cap S$.

2. $B \cup S$.

3. $\overline{(\overline{A} \cap B)} \cap S$.

9.1.6 In a survey, 100 people drink Coke© or Pepsi©, 70 drink Coke while 45 drink Pepsi. How many people drink both?

9.1.7 The Nielson ratings claim 232,679 people watch *The Brady Bunch* and *Andy Griffith* every day while 532,568 watch either program. If 392,101 people watch *Andy Griffith*, how many people watch *The Brady Bunch*?

9.1.8 In a survey at Emory University, of 200 students who say they drink beer or wine, 70 drink only beer while 45 drink only wine. How many people drink both beer and wine? Explain the difference between this problem and Exercise 9.1.6.

9.2   Standard Normal Distribution Table

Z	0.00	0.01	0.02	0.03	0.04	0.05	0.06	0.07	0.08	0.09
0.0	0.5000	0.5040	0.5080	0.5120	0.5160	0.5199	0.5239	0.5279	0.5319	0.5359
0.1	0.5398	0.5438	0.5478	0.5517	0.5557	0.5596	0.5636	0.5675	0.5714	0.5753
0.2	0.5793	0.5832	0.5871	0.5910	0.5948	0.5987	0.6026	0.6064	0.6103	0.6141
0.3	0.6179	0.6217	0.6255	0.6293	0.6331	0.6368	0.6406	0.6443	0.6480	0.6517
0.4	0.6554	0.6591	0.6628	0.6664	0.6700	0.6736	0.6772	0.6808	0.6844	0.6879
0.5	0.6915	0.6950	0.6985	0.7019	0.7054	0.7088	0.7123	0.7157	0.7190	0.7224
0.6	0.7257	0.7291	0.7324	0.7357	0.7389	0.7422	0.7454	0.7486	0.7517	0.7549
0.7	0.7580	0.7611	0.7642	0.7673	0.7704	0.7734	0.7764	0.7794	0.7823	0.7852
0.8	0.7881	0.7910	0.7939	0.7967	0.7995	0.8023	0.8051	0.8078	0.8106	0.8133
0.9	0.8159	0.8186	0.8212	0.8238	0.8264	0.8289	0.8315	0.8340	0.8365	0.8389
1.0	0.8413	0.8438	0.8461	0.8485	0.8508	0.8531	0.8554	0.8577	0.8599	0.8621
1.1	0.8643	0.8665	0.8686	0.8708	0.8729	0.8749	0.8770	0.8790	0.8810	0.8830
1.2	0.8849	0.8869	0.8888	0.8907	0.8925	0.8944	0.8962	0.8980	0.8997	0.9015
1.3	0.9032	0.9049	0.9066	0.9082	0.9099	0.9115	0.9131	0.9147	0.9162	0.9177
1.4	0.9192	0.9207	0.9222	0.9236	0.9251	0.9265	0.9279	0.9292	0.9306	0.9319
1.5	0.9332	0.9345	0.9357	0.9370	0.9382	0.9394	0.9406	0.9418	0.9429	0.9441
1.6	0.9452	0.9463	0.9474	0.9484	0.9495	0.9505	0.9515	0.9525	0.9535	0.9545
1.7	0.9554	0.9564	0.9573	0.9582	0.9591	0.9599	0.9608	0.9616	0.9625	0.9633
1.8	0.9641	0.9649	0.9656	0.9664	0.9671	0.9678	0.9686	0.9693	0.9699	0.9706
1.9	0.9713	0.9719	0.9726	0.9732	0.9738	0.9744	0.9750	0.9756	0.9761	0.9767
2.0	0.9772	0.9778	0.9783	0.9788	0.9793	0.9798	0.9803	0.9808	0.9812	0.9817
2.1	0.9821	0.9826	0.9830	0.9834	0.9838	0.9842	0.9846	0.9850	0.9854	0.9857
2.2	0.9861	0.9864	0.9868	0.9871	0.9875	0.9878	0.9881	0.9884	0.9887	0.9890
2.3	0.9893	0.9896	0.9898	0.9901	0.9904	0.9906	0.9909	0.9911	0.9913	0.9916
2.4	0.9918	0.9920	0.9922	0.9925	0.9927	0.9929	0.9931	0.9932	0.9934	0.9936
2.5	0.9938	0.9940	0.9941	0.9943	0.9945	0.9946	0.9948	0.9949	0.9951	0.9952
2.6	0.9953	0.9955	0.9956	0.9957	0.9959	0.9960	0.9961	0.9962	0.9963	0.9964
2.7	0.9965	0.9966	0.9967	0.9968	0.9969	0.9970	0.9971	0.9972	0.9973	0.9974
2.8	0.9974	0.9975	0.9976	0.9977	0.9977	0.9978	0.9979	0.9979	0.9980	0.9981
2.9	0.9981	0.9982	0.9982	0.9983	0.9984	0.9984	0.9985	0.9985	0.9986	0.9986
3.0	0.9987	0.9987	0.9987	0.9988	0.9988	0.9989	0.9989	0.9989	0.9990	0.9990

FIGURE 9.3:   Standard Normal Distribution: values correspond to area shown in figure.

9.3 Student's *t*-Distribution

df	.75	.80	.85	.90	.95	.975	.99	.995	.9975	.999
1	1.000	1.376	1.963	3.078	6.314	12.71	31.82	63.66	127.3	318.3
2	0.816	1.061	1.386	1.886	2.920	4.303	6.965	9.925	14.09	22.33
3	0.765	0.978	1.250	1.638	2.353	3.182	4.541	5.841	7.453	10.21
4	0.741	0.941	1.190	1.533	2.132	2.776	3.747	4.604	5.598	7.173
5	0.727	0.920	1.156	1.476	2.015	2.571	3.365	4.032	4.773	5.893
6	0.718	0.906	1.134	1.440	1.943	2.447	3.143	3.707	4.317	5.208
7	0.711	0.896	1.119	1.415	1.895	2.365	2.998	3.499	4.029	4.785
8	0.706	0.889	1.108	1.397	1.860	2.306	2.896	3.355	3.833	4.501
9	0.703	0.883	1.100	1.383	1.833	2.262	2.821	3.250	3.690	4.297
10	0.700	0.879	1.093	1.372	1.812	2.228	2.764	3.169	3.581	4.144
11	0.697	0.876	1.088	1.363	1.796	2.201	2.718	3.106	3.497	4.025
12	0.695	0.873	1.083	1.356	1.782	2.179	2.681	3.055	3.428	3.930
13	0.694	0.870	1.079	1.350	1.771	2.160	2.650	3.012	3.372	3.852
14	0.692	0.868	1.076	1.345	1.761	2.145	2.624	2.977	3.326	3.787
15	0.691	0.866	1.074	1.341	1.753	2.131	2.602	2.947	3.286	3.733
16	0.690	0.865	1.071	1.337	1.746	2.120	2.583	2.921	3.252	3.686
17	0.689	0.863	1.069	1.333	1.740	2.110	2.567	2.898	3.222	3.646
18	0.688	0.862	1.067	1.330	1.734	2.101	2.552	2.878	3.197	3.610
19	0.688	0.861	1.066	1.328	1.729	2.093	2.539	2.861	3.174	3.579
20	0.687	0.860	1.064	1.325	1.725	2.086	2.528	2.845	3.153	3.552
21	0.686	0.859	1.063	1.323	1.721	2.080	2.518	2.831	3.135	3.527
22	0.686	0.858	1.061	1.321	1.717	2.074	2.508	2.819	3.119	3.505
23	0.685	0.858	1.060	1.319	1.714	2.069	2.500	2.807	3.104	3.485
24	0.685	0.857	1.059	1.318	1.711	2.064	2.492	2.797	3.091	3.467
25	0.684	0.856	1.058	1.316	1.708	2.060	2.485	2.787	3.078	3.450
26	0.684	0.856	1.058	1.315	1.706	2.056	2.479	2.779	3.067	3.435
27	0.684	0.855	1.057	1.314	1.703	2.052	2.473	2.771	3.057	3.421
28	0.683	0.855	1.056	1.313	1.701	2.048	2.467	2.763	3.047	3.408
29	0.683	0.854	1.055	1.311	1.699	2.045	2.462	2.756	3.038	3.396

FIGURE 9.3: Student's *t* distribution (from [39]).

9.4   Solutions to Problems

Problem 2.2.1 The number of 5-card poker hands.

Hands	Number of Hands
straight flush	40
four of a kind	624
full house	3,744
flush	5,108
straight	10,200
three of a kind	54,912
two pair	123,552
one pair	1,098,240
high card	1,302,540

Problem 2.4.1

1st Roll	P(1st roll)	P(1st then pass)	P(1st then don't pass)
7 or 11	$\frac{8}{36}$	.2222	.0000
2, 3, 12	$\frac{4}{36}$	.0000	.1111
4	$\frac{3}{36}$	$\frac{3}{36}\frac{3}{9} = .0278$	$\frac{3}{36}\frac{6}{9} = .0556$
5	$\frac{4}{36}$	$\frac{4}{36}\frac{4}{10} = .0444$	$\frac{4}{36}\frac{6}{10} = .0667$
6	$\frac{5}{36}$	$\frac{5}{36}\frac{5}{11} = .0631$	.0757
8	$\frac{5}{36}$	.0631	.0757
9	$\frac{4}{36}$	.0444	.0667
10	$\frac{3}{36}$	.0278	.0556

Problem 2.7.1 $\frac{1}{5525}$.

Problem 4.2.1 The 8 of hearts.

Problem 4.2.2 The 7 of diamonds.

Problem 7.2.1 A 4×4 magic square.

Problem 7.6.1 Consider the labeling of $6 - purge$ and $L - purge$ as described in Figure 9.4.

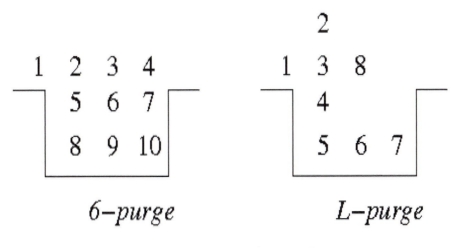

FIGURE 9.4: Labeling for Problem 7.6.1.

1. $6 - purge$: Suppose the left x (number 1 in Figure 9.4) is empty and the right x (number 3 in Figure 9.4) is filled. Moves: (1) 8 $\rightarrow$ 2; (2) 10 $\rightarrow$ 8; (3) 7 $\rightarrow$ 5; (4) 3 $\rightarrow$ 1; (5) 8 $\rightarrow$ 2; (6) 1 $\rightarrow$ 3.

2. $L - purge$: Suppose the left x (number 1 in Figure 9.4) is empty and the right x (number 8 in Figure 9.4) is filled. Moves: (1) 8 $\rightarrow$ 1; (2) 5 $\rightarrow$ 3; (3) 7 $\rightarrow$ 5; (4) 2 $\rightarrow$ 4; (5) 5 $\rightarrow$ 3; (6) 1 $\rightarrow$ 3.

Problem 7.6.2

(1)	(3)	(4)	(5)	(6)
(2,0) → (0,0)	(3,1) → (1,1)	(-1,3) → (-1,1)	(-3,1) → (-1,1)	(0,-1) → (-2,-1)
(1,-2) → (1,0)	(1,2) → (1,0)	(-2,1) → (0,1)	(-1,0) → (-1,2)	(-1,-3) → (-1,-1)
(0,0) → (2,0)	(3,-1) → (3,1)	(1,2) → (-1,2)	(-3,0) → (-1,0)	(1,-3) → (-1,-3)
(2)	(2,-1) → (2,1)	(1,3) → (-1,3)	(-1,-1) → (-1,1)	(-1,0) → (-1,-2)
(1,2) → (1,0)	(3,1) → (1,1)	(-1,3) → (-1,1)	(-3,-1) → (-1,-1)	(-1,-3) → (-1,-1)
(-1,1) → (1,1)	(1,0) → (1,2)	(0,1) → (-2,1)	(-1,2) → (-1,0)	(-2,-1) → (0,-1)
(1,0) → (1,2)				

Problem 8.3.1 1100101_2.

Problem 8.3.2 202022_3.

Problem 8.3.3 111423_5.

Problem 8.3.4 Remove exactly those chips added by your opponent. Otherwise play as in normal Nim.

Problem 8.3.5 Northcott's game: treat each row as a Nim column. If your opponent backs up, treat it as Poker Nim.

Problem 8.3.6 Another Nim variant. Treat space between coins as Nim piles.

Problem 8.3.7 Use the hint.

Problem 8.5.1 Use one red stick and one red then blue stick.

9.5   Selected Even Exercises

Chapter 1

Section 1.2

1.2.2: 12.

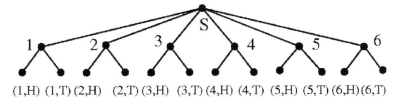

$(1,H)$ $(1,T)$ $(2,H)$ $(2,T)$ $(3,H)$ $(3,T)$ $(4,H)$ $(4,T)$ $(5,H)$ $(5,T)$ $(6,H)$ $(6,T)$

FIGURE 9.5:   The choice tree for Exercise 1.2.2.

1.2.4: all 52 cards in the deck.
1.2.6: 312.
1.2.8: 24.

Section 1.3

1.3.2: $P(\text{spade}) = \frac{1}{4}$; $P(\text{ace}) = \frac{1}{13}$.
1.3.4: Sample size $= 104$; $P(\text{single element of sample space}) = \frac{1}{104}$.
1.3.6: $P(\Diamond \text{ not J, Q, K }) = \frac{5}{26}$.
1.3.8: $P(6 \text{ and } \clubsuit) = \frac{1}{24}$.

Section 1.4

1.4.2: $P(8 \text{ and } \heartsuit) = \frac{1}{52}$.
1.4.4: $P(\text{sum of } 4) = \frac{1}{12}$.
1.4.6: $P(\text{sum of 8 or doubles}) = \frac{5}{18}$.
1.4.8: (1) $P(\text{two heads}) = \frac{103}{400}$; (2) $P(\text{two tails}) = \frac{99}{400}$.
1.4.10: (1) $P(\text{superbowl ticket}) = \frac{9}{20}$; (2) $P(\text{bowl II} \mid \text{superbowl ticket})$ $= \frac{2}{3}$.
1.4.12: $P(\text{attended all camps} \mid \text{first round productive season}) = 0.6989$.

Section 1.5

1.5.2: 52!
1.5.4: 22,100.

1.5.6: 12!
1.5.8: 120.
1.5.10: 200.
1.5.12: (1) 3!4!2!4!; (2) (6912)(4!) = 165888; (3) (6912)(4) = 27648.

Section 1.6

1.6.2: 7.
1.6.4: $0.25.
1.6.6: -$0.125.
1.6.8: $p^2 - 3p + 3$.

Section 1.7

1.7.2: P(longer than 3 tosses) = 1/8; P(longer than 4 tosses) = 1/16.
1.7.4: Tom should get 3/4 of the prize and Jane should get 1/4 of the prize.
1.7.6: Tom should get 11/16 of the prize and Jane should get 5/16 of the prize.
1.7.8: Tom should get 5/16 of the prize and Jane should get 11/16 of the prize.
1.7.10: No.

Section 1.8

1.8.2: $\frac{P(E)}{1-P(E)}$.
1.8.4: 215:1.
1.8.6: 10:3.
1.8.8: 5:1.
1.8.10: 16:1.

Section 1.9

1.9.2: 2.5; 1.25; 1.12.
1.9.4: 7; 14.
1.9.6: 6.188; 2.487.

Chapter 2

Section 2.2

2.2.2: 0.002.
2.2.4: 22100.
2.2.6: 0.1111.
2.2.8: 0.0851.

Section 2.3

2.3.2: -0.027.

2.3.4: In European Roulette: EV(red and 2) = -0.0541;
In American Roulette: EV(red and 2) = -0.1053.

2.3.6: There are 38 numbers on the Roulette wheel. There are 6 numbers in any two rows, so true odds of a bet on any two rows are 32:6. There are 4 numbers in any 4-number square, so true odds of a bet on any 4-number square are 34:4. Similarly, there are 12 choices in any 12-number combinations, so true odds of a bet on any 12-number combinations are 26:12.

2.3.8: EV(two adjacent numbers) = -0.0526.

Section 2.4

2.4.2:

1st Roll	P(1st Roll)	P(1st then Pass)	P(1st then don't Pass)
6	$\frac{5}{36}$	$\frac{5}{36} \cdot \frac{5}{11} = 0.0631$	$\frac{5}{36} \cdot \frac{6}{11} = 0.0758$
8	$\frac{5}{36}$	$\frac{5}{36} \cdot \frac{5}{11} = 0.0631$	$\frac{5}{36} \cdot \frac{6}{11} = 0.0758$
9	$\frac{4}{36}$	$\frac{4}{36} \cdot \frac{4}{10} = 0.0444$	$\frac{4}{36} \cdot \frac{6}{10} = 0.0667$
10	$\frac{3}{36}$	$\frac{3}{36} \cdot \frac{3}{9} = 0.0278$	$\frac{3}{36} \cdot \frac{6}{9} = 0.0556$

2.4.4:

Place Bets	4	5	6	8	10	12
True Odds	33:3	32:4	31:5	31:5	32:4	33:3

2.4.6: 10:1.

2.4.8: The house advantage on a hard 8 bet, a hard 10 bet and a hard 12 bet is just the expected value of loss for the player. Now the true odds for a hard 8 bet, a hard 10 bet and a hard 12 bet are 10:1, 8:1 and 35:1, respectively. Hence, using the payouts from Table 2.5, the house advantage (house edge) for each bet is calculated as follows:

EV(hard 8 bet) = $-9(\frac{1}{11}) + 1(\frac{10}{11}) = 0.091$

EV(hard 10 bet) = $-7(\frac{1}{9}) + 1(\frac{8}{9}) = 0.111$

EV(hard 12 bet) = $-29(\frac{1}{36}) + 1(\frac{35}{36}) = 0.167$.

2.4.10: -0.0135.

2.4.12: True odds for a Big 6 bet are 6:5. So EV(Big 6) = $-1(\frac{5}{11}) + 1(\frac{6}{11})$ = 0.091.

Section 2.5

2.5.2: If Monty Hall does not know where the car is, then he can open the door which has a car behind it and the game will end. Now $P(C_i) = \frac{1}{3}$ for $i = 1, 2, 3$. The following table gives the conditional probabilities. The entry in row 2, column 2 of the table gives the $P(H_{12}|C_1)$, the entry in row 6, column 3 of the table gives the $P(H_{31}|C_2)$, etc.

	C_1	C_2	C_3
H_{12}	$\frac{1}{6}$	$\frac{1}{6}$	$\frac{1}{6}$
H_{13}	$\frac{1}{6}$	$\frac{1}{6}$	$\frac{1}{6}$
H_{21}	$\frac{1}{6}$	$\frac{1}{6}$	$\frac{1}{6}$
H_{23}	$\frac{1}{6}$	$\frac{1}{6}$	$\frac{1}{6}$
H_{31}	$\frac{1}{6}$	$\frac{1}{6}$	$\frac{1}{6}$
H_{32}	$\frac{1}{6}$	$\frac{1}{6}$	$\frac{1}{6}$

2.5.4: $\frac{9}{14}$.

Section 2.6

2.6.2: -0.0787.

2.6.4: -0.3442.

2.6.6: 0.0008.

Section 2.7

2.7.2: -1.

2.7.4: 0.4191.

2.7.6: 0.0006.

2.7.8: 0.0664.

2.7.10: $330.13.

2.7.12: $EV(\text{surrender}) = 3p - 2$; $EV(\text{not surrender}) = 4p - 2$.

2.7.14: 0.0322.

Section 2.8

2.8.2:

No. of Opponent's safe points	0	1	2	3	4	5	6
Probability of entering on next roll	1	$\frac{35}{36}$	$\frac{8}{9}$	$\frac{3}{4}$	$\frac{5}{9}$	$\frac{11}{36}$	0

2.8.4: $\frac{49}{6}$.

2.8.6: $\frac{13}{18}$.

2.8.8: $P(\text{win on first roll}) = \frac{5}{36}$. So if the opponent does not win on the second roll, then double the stake.

2.8.10: $\frac{4}{9}$.

2.8.12: $P(\text{win on first roll}) = \frac{17}{36}$. So you should not double the stake. If you don't win, then you should not accept double from your opponent on second roll.

Chapter 3

Section 3.2

3.2.2: $EV(\text{spades}) = 1.25$; $var(\text{spades}) = 2.5$.

3.2.4: (1) 0.2592; (2) 0.3456; (3) 0.9898.

3.2.6: (1) 0.4004; (2) 0.284; (3) 0.00096.

3.2.8: 0.4697.

3.2.10: (1) 0.3355; (2) 0.2031.

3.2.12: The score is more likely to be 3-1. The score is more likely to be 4-2.

Section 3.3

3.3.2: 0.4602.

3.3.4: 0.5901.

3.3.6: 1.

3.3.8: 0.7357.

3.3.10: 0.6255.

3.3.12: 0.8186.

Section 3.4

3.4.2: (1) 0.0025; (2) 0.015; (3) 0.0446; (4) 0.0892; (5) 0.9975.

3.4.4: (1) 0.3679; (2) 0.1839; (3) 0.3679; (4) 0.6321.

3.4.6: (1) 0.0067; (2) 0.0337; (3) 0.1246.

Section 3.5

3.5.2: (1) 0.0538; (2) 0.0125; (3) 0.0027; (4) 1; (5) 0.7983.

3.5.4: (1) 0.1936; (2) 0.0308; (3) 0.0030; (4) 0.0165.

Section 3.6

3.6.2: $m; $m.

3.6.4: $20, $30, $45.

Section 3.7

3.7.2: As t grows larger with respect to m, $\left(\frac{0.52}{0.48}\right)^{m-t}$ can be seen to grow close to 0.

Chapter 4

Section 4.2

4.2.2: 6♡, 4♣, 4♢, 4♠; face down card: 7♡.
4.2.4: 4♣, 5♢, joker, 6♡; face down card: 7♠.
4.2.6: 4.
4.2.8: Label the 8 cards in the deck 1, 2, 3, 4, 5, 6, 7, 8.

hand dealt	$n-1$ ordering matched to hand	face down card	hand dealt	$n-1$ ordering matched to hand	face down card
123	12	3	134	13	4
145	14	5	156	15	6
167	16	7	178	17	8
182	18	2	214	21	4
235	23	5	246	24	6
256	25	6	267	26	7
271	27	1	286	28	6
315	31	5	326	32	6
348	34	8	358	35	8
361	36	1	372	37	2
382	38	2	416	41	6
427	42	7	432	43	2
453	45	3	468	46	8
471	47	1	485	48	5
517	51	7	521	52	1
537	53	7	542	54	2
563	56	3	574	57	4
586	58	6	618	61	8
621	62	1	637	63	7
643	64	3	654	65	4
675	67	5	687	68	7
713	71	3	725	72	5
734	73	4	748	74	8
758	75	8	764	76	4
782	78	2	814	81	4
825	82	5	831	83	1
842	84	2	851	85	1
863	86	3	873	87	3

Section 4.3

4.3.2: 0.625.

4.3.4: 0.6301.

4.3.6: 485.

Section 4.4

4.4.2: The 9-card trick works even if the victim lies to you about the card since any number he spells out moves the card third from the bottom (as any number from 1-9 is spelled with at least three letters). Then when he spells *of*, it moves the card fifth from the bottom (in the middle). Then when he spells out suit of the card, it keeps the card in the middle (as any suit is spelled with at least five letters).

4.4.4: Answers may vary.

Section 4.5

4.5.2: If Al kills either Bob or Carl on the first shot, then P(Al winning) $= 0$. If Al misses on the first shot, then P(Al winning) $= \frac{1}{2}$.

Chapter 5

Section 5.2

5.2.2: Bob; Bob.

5.2.4: Al; Al.

5.2.6: Bob had a better average when left-handed pitching, right-handed pitching, day, and night game are considered separately. Whereas Al had a better average when left-/right-handed pitching was combined with a day/night game. Also, Al had overall better average.

Section 5.3

5.3.2: 79.9.

5.3.4: Answers may vary.

5.3.6: NCAA rating $= \frac{[(8.4 \cdot yards) + (330 \cdot T) - (200 \cdot I) + (100 \cdot C)]}{attempts}$.

5.3.8: Answers may vary.

Section 5.4

5.4.2: Robertson scored higher than West.

5.4.4: The comparison stem plot looks almost like American stem plot as leading averages in 1920s are higher than 1980s.

1920s					1980s		
			31	3			
			32	3	4		
			33	1	4	6	
			34	0			
		3	35	1	3		
			36				
		0	37	0			
7	4	0	38				
	8	7	39				
	3	1	40				
			41				
		4	42				

5.4.6:

	first quartile	second quartile	third quartile
1920s	0.380	0.392	0.403
1980s	0.324	0.335	0.351

5.4.14: mean = 39.9; variance = 66.27; standard deviation = 8.14.

Section 5.5

5.5.2: (1) 0.54; (2) [0.509, 0.571].

5.5.4: (1) 0.167; (2) [0.088, 0.246].

Section 5.6

5.6.2: 75%; 89%; 95%.

5.6.4: Runner 1, with weight of 178 pounds, is more overweight.

5.6.6: 86%.

5.6.8: 89%.

5.6.10: 89%; 82%.

5.6.12: Ward brought approximately 9 more wins to the Carolina Hurricanes in 2007-2008 season than an average goalie on his team.

Chapter 6

Section 6.2

6.2.2: Answers may vary.

Section 6.3

6.3.2:

1. The claim is $\mu = 0.65$, so the hypotheses are $\mathcal{H}_0 : \mu = 0.65$ and $\mathcal{H}_a : \mu \neq 0.65$ with level of significance $\alpha = 0.05$. The test value is $Z = \frac{0.68-0.65}{0.05/\sqrt{36}} = 3.6$. The critical region is $Z < -1.96$ or $Z > 1.96$. Since the test value is in the critical region, reject the null hypothesis $\mathcal{H}_0$ in favor of the alternative hypothesis $\mathcal{H}_a$.

2. Now the test value is $Z = \frac{0.68-0.65}{0.05/\sqrt{100}} = 6$. The critical region is still $Z < -1.96$ or $Z > 1.96$. Since the test value is in the critical region, reject the null hypothesis $\mathcal{H}_0$ in favor of the alternative hypothesis $\mathcal{H}_a$.

3. P(type I error) $= 0.05$.

6.3.4: The claim is $\mu > 3$, so the hypotheses are $\mathcal{H}_0 : \mu = 3$ and $\mathcal{H}_a : \mu > 3$ with level of significance $\alpha = 0.02$. The test value is $Z = \frac{3.2-3}{0.25/\sqrt{49}} = 5.6$. The critical region is $Z > 2.055$. Since the test value is in the critical region, reject the null hypothesis $\mathcal{H}_0$ in favor of the alternative hypothesis $\mathcal{H}_a$.

6.3.6: The claim is $\mu = 20$, so the hypotheses are $\mathcal{H}_0 : \mu = 20$ and $\mathcal{H}_a : \mu \neq 20$ with level of significance $\alpha = 0.02$. The test value is $t = \frac{17.2-20}{4.44/\sqrt{10}} = -1.99$. The critical region is $t < -2.821$ or $t > 2.821$ for $df = 9$. Since the test value is not in the critical region, accept the null hypothesis $\mathcal{H}_0$.

Section 6.5

6.5.2: The correlation coefficient $= 0.9026$; the least squares approximating line is $y = 5363.922x - 1024.258$.

6.5.4: The correlation coefficient $= 0.9074$; the least squares approximating line is $y = 2759.244x - 405.410$.

6.5.6: The correlation coefficient $= 0.9479$.

6.5.8: The correlation coefficient $= -0.0881$.

Section 6.6

6.6.2: Trammel$_{RBA} = \frac{0.300}{0.269} \times 100 = 111.52$.

6.6.4: Yastrzemski$_{MABA} = \frac{0.301}{0.230} \times 0.255 = 0.334$.

6.6.6: Rivers$_{MABA} = \frac{0.333}{0.269} \times 0.255 = 0.316$.

Section 6.7

6.7.2: The claim is $\mu_W = \mu_R$, so the hypotheses are $\mathcal{H}_0 : \mu_W = \mu_R$ and $\mathcal{H}_a : \mu_W \neq \mu_R$ with level of significance $\alpha = 0.05$. The t-statistic is $t = \dfrac{1646.5 - 1589}{\sqrt{\frac{(506.51)^2}{8} + \frac{(515.33)^2}{8}}} = 0.2251$. The critical region is $t \leq -2.365$ or $t \geq 2.365$ for $df = 7$. Since the test value is not in the critical region, accept the null hypothesis $\mathcal{H}_0$.

6.7.4: The claim is $\mu_W = \mu_R$, so the hypotheses are $\mathcal{H}_0 : \mu_W = \mu_R$ and $\mathcal{H}_a : \mu_W \neq \mu_R$ with level of significance $\alpha = 0.02$. The t-statistic is $t = \dfrac{1646.5 - 1589}{\sqrt{\frac{(506.51)^2}{8} + \frac{(515.33)^2}{8}}} = 0.2251$. The critical region is $t \leq -2.998$ or $t \geq 2.998$ for $df = 7$. Since the test value is not in the critical region, accept the null hypothesis $\mathcal{H}_0$.

6.7.6: Answers may vary.

6.7.8: Answers may vary.

Chapter 7

Section 7.2

7.2.2:

2	9	4
7	5	3
6	1	8

7.2.4:

1	14	8	11
15	4	10	5
12	7	13	2
6	9	6	16

7.2.6:

11	24	7	20	3
4	12	25	8	16
17	5	13	21	9
10	18	1	14	22
23	6	19	2	15

7.2.8:

1	5	6	2	4	3
4	3	2	5	6	1
2	4	3	1	5	6
5	6	1	4	3	2
3	2	5	6	1	4
6	1	4	3	2	5

7.2.10:

3	7	9	4	1	8	5	2	6
8	5	4	2	6	7	1	3	9
1	6	2	3	9	5	8	7	4
2	1	6	9	3	4	7	5	8
9	8	5	1	7	6	2	4	3
4	3	7	5	8	2	9	6	1
6	2	3	8	5	9	4	1	7
7	4	8	6	2	1	3	9	5
5	9	1	7	4	3	6	8	2

Section 7.3

7.3.2: Yes.

7.3.4: (1) Move disk 1 to peg 2; (2) move disk 1 to peg 3; (3) move disk 2 to peg 2; (4) move disk 1 to peg 1; (5) move disk 2 to peg 3; (6) move disk 1 to peg 3.

7.3.6: Three white (black) disks are labeled W1(B1), W2(B2), W3(B3), with W1(B1) being the smallest and W3(B3) being the largest disk. Peg 1 contains the disks W3, B2, W1; peg 2 contains the disks B3, W2, B1; and peg 3 is empty. There are 10 moves involved and this number is optimal. Here is the list of moves: (1) move W1 to peg 2; (2) move B2 to peg 3; (3) move W1 to peg 3; (4) move B1 to peg 3; (5) move W2 to peg 1; (6) move B1 to peg 2; (7) move W1 to peg 1; (8) move B1 to peg 1; (9) move B2 to peg 2; (10) move B1 to peg 2.

7.3.8: See Figure 7.7. Let W1(B1) be the smallest white (black) disk and W2(B2) be the largest white (black) disk. (1) Move B1 to peg 2; (2) move W2 to peg 3; (3) move B1 to peg 1; (4) move W1 to peg 3; (5) move B1 to peg 3; (6) move B2 to peg 1; (7) move B1 to peg 1.

Section 7.4

7.4.2: Rotate the cubes to place them in standard positions as illustrated in Figure 7.9 with color coordination described in Figure 9.6. Then place Cube 2 in the bottom, Cube 1 on top of the Cube 2, Cube 3 on top of Cube 1, and finally Cube 4 on top of Cube 3.

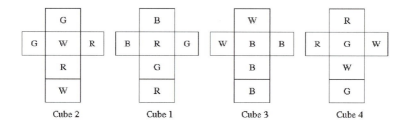

FIGURE 9.6:   The cubes of the solution for Exercise 7.4.2.

7.4.4: See Figure 9.6.

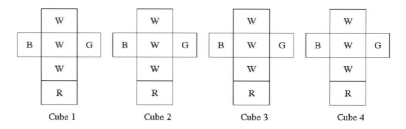

FIGURE 9.7:   The cubes of the solution for Exercise 7.4.4.

7.4.6: No. Take all white, all green, all blue, all red cubes.
7.4.8: 41,472.

Section 7.5

7.5.2: (1) Press the (1,1) light; (2) press the (2,2) light; (3) press the (1,2) light; (4) press the (2,1) light.

7.5.4: (1) Press 1; (2) press 2; (3) press 3. Now you start to press 1, 2, 3 again which is same as not pressing.

7.5.6: (1) Press 1; (2) press 2; (3) press 3; (4) press 4. Now you repeat all over again as we have pressed all the lights. No, the result would not change if the path was longer.

7.5.8: No.

Section 7.6

7.6.2: The possible positions for a final single peg is all the z positions in Figure 7.20. This list can be reduced to the positions $(0, -1)$, $(3, -1)$, $(-3, -1)$, $(1, 3)$, and $(1, -3)$.

7.6.4: If there is a one-peg solution and the two open holes are labeled x and y, then the final peg must be on one of the positions labeled z in Figure 7.20. If the two open holes are both labeled z, then there is no one-peg solution.

7.6.6: The possible positions for a final single-peg solution are all the positions labeled z in Figure 7.26. This list cannot be reduced.

7.6.8: $F - O, M - F, K - M, N - L, D - M, A - D, G - B, B - I, L - N, C - J, O - F, F - M, M - O$.

7.6.10:

(1)	(3)	(4)	(5)	(6)
(1,1) → (-1,1)	(1,3) → (1,1)	(3,-1) → (1,-1)	(-2,-1) → (0,-1)	(-2,-1) → (0,-1)
(0,-1) → (0,1)	(-1,3) → (1,3)	(2,1) → (2,-1)	(-1,-3) → (-1,-1)	(-2,1) → (-2,-1)
(-1,1) → (1,1)	(-1,2) → (1,2)	(3,1) → (3,-1)	(0,-1) → (-2,-1)	(-3,-1) → (-1,-1)
(2)	(2,1) → (0,1)	(1,-2) → (1,0)	(1,-2) → (-1,-2)	(0,-1) → (-2,-1)
(2,1) → (0,1)	(1,3) → (1,1)	(3,-1) → (1,-1)	(1,-3) → (-1,-3)	(-3,1) → (-3,-1)
(1,-1) → (1,1)	(0,1) → (2,1)	(1,0) → (1,-2)	(-1,3) → (-1,-1)	(-3,-1) → (-1,-1)
(0,1) → (2,1)				

Chapter 8

Section 8.2

8.2.2: **P**-positions: 0, 6, 12, 18, 24, 30, 36; **N**-positions: rest.

8.2.4: (1) 0, 3, 6, 9, 12, 15, 18, 21, 24, 27, 30, 33;

(2) 0, 2, 4, 6, 8, 10, 12, 14, 16, 18, 20, 22, 24, 26, 28, 30, 32;

(3) 1.

Section 8.3

8.3.2: Move 3 chips from the pile containing 7 chips.

8.3.4: Move all the chips from the fourth pile containing 8 chips.

8.3.6: (1) $\langle 53, 85, 102, 56, 127 \rangle_1 \rightarrow \langle 53, 85, 74, 42, 127 \rangle_2$;

(2) $\langle 53, 85, 102, 56, 127 \rangle_1 \rightarrow \langle 53, 29, 45, 56, 61 \rangle_2$;

(3) $\langle 53, 85, 102, 56, 127 \rangle_1 \rightarrow \langle 53, 53, 53, 53, 53 \rangle_2$.

8.3.8: 12.

8.3.10: Start with an edge in the middle, between vertex 4 and 5. Then add an edge symmetric to player 2's edge from the previous step. Strategy remains the same for 10 vertices, so nothing changes.

Section 8.4

8.4.2:

x	0	1	2	3	4	5	6	7	8	9	10	11
sg(x)	0	1	2	2	3	3	0	0	0	0	0	4

8.4.4: $sg(x) = 0$ for all even x.

x	1	3	5	7	9	11	13	15	17	19	21
sg(x)	0	1	2	3	0	1	2	3	0	1	2

8.4.6: If you start with even number of chips, then Sprague-Grundy function values are as follows:

x	0	1	2	3	4	5	6	7	8	9	10 ...
sg(x)	0	0	1	0	2	0	3	0	4	0	5 ...

If you start with odd number of chips, then Sprague-Grundy function values are as follows:

x	0	1	2	3	4	5	6	7	8	9	10	11 ...
sg(x)	0	1	0	0	0	2	0	3	0	4	0	5 ...

8.4.8: The following table gives Sprague-Grundy function values for each square of a standard 8×8 chessboard for Wythoff's game with a rook starting in the upper right position of the board.

7	6	5	4	3	2	1	0
6	7	4	5	2	3	0	1
5	4	7	6	1	0	3	2
4	5	6	7	0	1	2	3
3	2	1	0	7	6	5	4
2	3	0	1	6	7	4	5
1	0	3	2	5	4	7	6
0	1	2	3	4	5	6	7

8.4.10: (1) If second pin has already been knocked down and there are 13 pins, then it is now a Nim game with two piles of 1 and 11 chips. So the Nim sum is 1010_2. Hence this is an **N**-position. (2) Knock down 7 and 8 pins together in the next move.

8.4.12: (1) If you start with odd number of chips and n is the number of chips, then $sg(n) = 1$ and $sg(x) = 0$ for all $x \neq n$. If you start with even number of chips, then Sprague-Grundy function values are as follows:

x	0	1	2	3	4	5	6	7	8	9	10	11	12 ...
sg(x)	0	0	0	0	1	0	2	0	3	0	4	0	5 ...

(2)

x	0	1	2	3	4	5	6	7	8	9	10	11	12	13	14	15
sg(x)	0	0	0	0	1	0	2	2	3	0	4	1	0	4	0	3
x	16	17	18	19	20	21	22	23	24	25	26	27	28	29	30	
sg(x)	0	1	0	0	0	3	0	2	1	5	2	0	3	0	4	

Section 8.5

8.5.2: (a) $1\frac{1}{16}$; (b) $-2\frac{13}{16}$; (c) $\frac{3}{4}$; (d) -3.

Section 8.6

8.6.2: Move the top branch with 2 leaves in the middle from the left tree.

8.6.4: Answers may vary.

Section 8.7

8.7.2: 1.5.

8.7.4: Let $G = \{G^L, G^R\}$ and $0 = \{0^L, 0^R\}$. Then $G+0 = G$; $G^L + 0^L = G^L$; and $0^R + G^R = G^R$. So $G + 0 = \{G^L, G^R\} = G$.

8.7.6: Let $G = \{G^L, G^R\}$. Then $-G = \{-G^L, -G^R\}$. Hence, $-(-G) = \{-(-G^L), -(-G^R)\} = \{G^L, G^R\} = G$.

Section 8.8

8.8.2: *26.

8.8.4: 8.

8.8.6: 1.

Chapter 9

Section 9.1

9.1.2: Another way of describing the last set (set in 5) is $\bar{A} \cup \bar{B} \cup \bar{C}$ (see Figure 9.8).

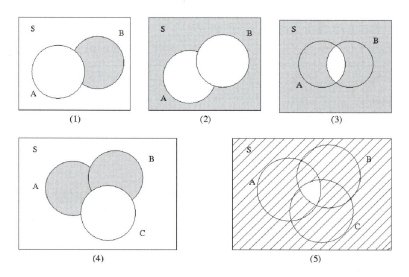

FIGURE 9.8:   Venn diagrams for Exercise A.1.2.

9.1.4: (1) $|H \cup B| = 40 + 45 = 85$; (2) $|H \cup B| = 40 + 45 - 15 = 70$.

9.1.6: $|C \cap P| = 70 + 45 - 100 = 15$.

9.1.8: $|B \cup W| = |B| + |W| + |B \cap W|$. So $|B \cap W| = |B \cup W| - |B| - |W| = 200 - 70 - 45 = 85$.

References

[1] Albert, J., *Teaching Statistics Using Baseball*, MAA, Washington, D.C., 2003.

[2] Albert, J. and J. Bennett, *Curve Ball*, Copernicus Books, New York, 2001.

[3] Anderson, M. and T. Feil, Turning lights out with linear algebra, *Mathematics Magazine*, Vol. 71, No. 4 (Oct. 1998), 300-303.

[4] Bennett, J., Ed. *Statistics in Sport*, Arnold, London, 1998.

[5] Berlekamp, E. R., The Hackenbush number system for the compression of numerical data, *Inform. and Control*, 26 (1974), 134-140.

[6] Berlekamp, E.R, J.H. Conway and R.K. Guy, *Winning Ways for Your Mathematical Plays*, Vol. I, Second Edition, A.K. Peters Ltd., Natick, MA, 2001.

[7] Berlekamp, E.R., J.H. Conway and R.K. Guy, *Winning Ways for Your Mathematical Plays*, Vol. II, Academic Press, San Diego, CA, 1998.

[8] Bewersdorff, J. (translated by David Kramer), *Luck, Logic & White Lies*, A.K. Peters Ltd., Wellesley, MA, 2005.

[9] Bialostocki, A., An application of elementary group theory to central solitaire, *The College Mathematics Journal*, Vol. 29, No. 3 (May 1998), 208-212.

[10] Boulton, C.L., A game with a complete mathematical theory, *The Annals of Mathematics*, Second Series, Vol. 3, No. 1/4 (1901-1902), 35-39.

[11] Chen, B. and J. Ankenman, *The Mathematics of Poker*, ConJelCo LLC, Pittsburgh, PA, 2006.

[12] Conway, J.H., *On Numbers and Games*, Second Edition, A.K. Peters Ltd., Natick, MA, 2001.

[13] Epstein, R.A., *The Theory of Gambling and Statistical Logic*, Academic Press, San Diego, CA, 1995.

[14] Freedman, D., R. Pisani and R. Purves, *Statistics*, Third Edition, W.W. Norton & Company, New York, 1998.

[15] Gardner, M., *Mathematical Puzzles and Diversions*, Simon and Schuster, New York, 1959.

[16] Haigh, J., *Taking Chances*, Oxford University Press, New York, 2003.

[17] Hall, P., On representation of subsets, *J. Lond. Math. Soc.* 10 (1935), 26-30.

[18] Hogg, R. and A. Craig, *Introduction to Mathematical Statistics*, Third Edition, Macmillan, New York, 1970.

[19] Instant Insanity, *Wikipedia, The Free Encyclopedia*, (May 2008), http://en.wikipedia.org/wiki/Instant_Insanity.

[20] Levitt, D., Empirical Analysis of Bunting, (October 2008) http://baseballanalysts.com/archives/2006/07/empirical_analy_1.php.

[21] Magic Squares, *Wikipedia, The Free Encyclopedia*, (June 2008), http:/en.wikipedia.org/wiki/Magic_square.

[22] Magriel, P., *Backgammon*, New York Times Book Co., New York, 1973.

[23] Major League Baseball Stats, http://espn.go.com.

[24] Martingale (probability theory), *Wikipedia, The Free Encyclopedia*, (May 2008) http://en.wikipedia.org/wiki/Martingale.

[25] Missigman, J. and Weida, R., An easy solution to mini lights out, *Mathematics Magazine*, Vol. 74, No. 1(Feb. 2001), 57-59.

[26] The Monty Hall Problem, *The Let's Make A Deal Web Page*, (April 2008), http://www.Letsmakeadeal.com.

[27] The Monty Hall Problem, *Wikipedia, The Free Encyclopedia*, (April 2008), http://en.wikipedia.org/wiki/Monty_Hall_problem.

[28] Moore, E. H., A generalization of the game of Nim, *Ann. Math.*, Vol. 11 (1910), 93-94.

[29] Mulcahy, C., Fitch Cheney's Five Card Trick, *Math. Horizons*, February 2003, 10-13.

[30] Nowakowski, R., Ed. *Games of No Chance*, Cambridge University Press, Cambridge, U.K., 1996.

[31] Oliver, T., *Kings of the Mound, A Pitcher's Rating Manual*, Los Angeles, self-published, 1944.

[32] Packel, E., *The Mathematics of Games and Gambling*, The Mathematical Association of America, Washington, D.C., 1981.

[33] Passer Rating, *Wikipedia, The Free Encyclopedia*, http://en.wikipedia.org/wiki/Passer_rating.

[34] Peg Solitaire, *Wikipedia, The Free Encyclopedia*, (November 2008), http://en.wikipedia.org/wiki/Peg_solitaire.

[35] Schell, M. J., *Baseball's All-Time Best Hitters - How Statistics Can Level the Playing Field*, Princeton University Press, Princeton, NJ, 1999.

[36] Schwartz, D.G., *Roll the Bones, The History of Gambling*, Gotham Books, Penguin Group, New York, 2006.

[37] Schwenk, Allen, personal communication, 1999.

[38] Simpson's Paradox, *Wikipedia, The Free Encyclopedia*, (May 2008), http://en.wikipedia.org/wiki/Simpson's_paradox.

[39] Student's *t*-Distribution, *Wikipedia, The Free Encyclopedia*, (September 2008), http://en.wikipedia.org/wiki/Student's_t-distribution.

[40] Sudoku, *Wikipedia, The Free Encyclopedia*, (September 2008), http://en.wikipedia.org/wiki/Sudoku.

[41] Sutner, K., Linear cellular atomata and the Garden-of-Eden, *Math. Intelligencer*, 11 (1989), 49-53.

[42] Thorp, E.O., *The Mathematics of Gambling*, Lyle Stuart, Secaucus, NJ, 1984.

[43] The Tower of Hanoi, *Wikipedia, The Free Encyclopedia*, (April 2008), http://en.wikipedia.org/wiki/Tower_of_Hanoi.

Index

k-tuple, 11
k-tuples, 11
t-statistic, 209
$\mathbf{N}$-positions, 278
$\mathbf{P}$-position, 278
6-purge, 271

Aaron, Hank, 185
additive identity, 268
additive inverse, 268
Ahmad al-Buni, 234
Albert, Jim, 159
alternative hypothesis, 205
ante, 72
anti-martingale, 126
associative, 269

backgammon, 79
batting average, 161
Bennett, Jay, 159
best fit line, 178
betting strategies, 124
Bicolored Tower of Hanoi, 251
Big 6, 64
Big 8, 64
bijection, 142
binary, 282
binomial coefficient, 94
binomial coefficient, $\binom{n}{k}$, 25
binomial experiment, 105
Binomial Theorem, 98
bipartite graph, 144
blackjack, 77
blot, 80

Blyth, Colin, 163

cancellation system, 127
Cardano, Girolamo, 2
cardinality, 144
central game, 267
Central Limit Theorem, 112
children, 292
choice tree, 6
Chuck-a-Luck, 69
coefficient of variation, 194
combination, 25
commutative, 269
complement, 320
complementary square, 236
complete information, 277
conditional expectation, 45
confidence interval, 186
counting, 242
craps, 58
    casino, 61
    street, 60
critical region, 206
critical values, 206
cross-hatching, 242

da Vinci, 4
de Fermat, Pierre, 4
de Méré, Chevalier, 3
degree, 144
degree of freedom, 210
dependent events, 16
derangement, 148, 253
discrete, 41

doubling cube, 81

edge, 143
elements, 319
ellipse, 239
empty set, ∅, 319
error, Type I, 207
error, Type II, 207
exclusive or, 285
expected value, 29

factorial, 24
fair game, 30, 301
field, 64
finite set, 319
five-card trick, 135
fixed point, 148
followers, 292
Fra Luca Paccioli, 4
free odds, 62
friends card trick, 152
full house, 50

Galileo, 2
gambler's ruin problem, 129
game, 310
geometric progression, 124
geometric series, 36
Gombaud, Antoine, 3
Gosset, William Sealy, 209
Granderson, Curtis, 191
graph, 143
Gray Code, 249

Hackenbush, 298
Hackenbush strings, 301
Hackenbush, Blue-Red, 298
Hackenbush, Green, 306
Hall, Monty, 66
hard 21, 77
hard bet, 65

histogram, 103
home, 80
house odds, 54
Howard Gams, 242
Howe, Gordie, 185
hypothesis test, 205

image, $Im(X)$, 42
impartial game, 278, 296
Inclusion-Exclusion Principle, 148
independent events, 5
infinite set, 319
inner table, 80
Instant Insanity, 255
insurance, 77
intersection, 320

James, Bill, 159

Keno, 74
Khabibulin, Nikolai, 197
Kopett, Leonard, 159

Laskar, Emanuel, 295
Latin Square, 240, 242
least squares regression line, 178
Lee, Cliff, 197
Let's Make A Deal, 66
lexicographic ordering, 137
Lights Out, 262
Lindsey, George, 159
linear correlation, 181
linear correlation coefficient, 181
Loh-Shu scroll, 234
loop, 256

magic constant, 233
magic square, 233
Manning, Peyton, 167
Marino, Dan, 168
martingale strategy, 124

matrix, 263
mean, μ, 41
mean-adjusted batting average, 226
Meche, Gil, 196
Merlin, 262
Mini Lights Out, 265
Monty Hall Problem, 66
Moore's Nim, 287
Moore, E. H., 287
Mosteller, Frederick, 159
Multiplication Principle, 6
mutually exclusive events, 14

natural numbers, 293
Nim, 280
Nim sum, 284
Nim_k, 287
Nimatron, 280
nimbers, 315
nine-card trick, 154
non-qualifying, 72
nopass line, 60
Northcott's Game, 289
null hypothesis, 205
Number Place, 242

Observed situational effect, 216
odds, 40
odds for, 41
Oliver, Ted, 196
one-to-one, 142
onto, 142
order of a set, 321
ordered pairs, 6
ordered triples, 11
orthogonal Latin Squares, 241
out neighbors, 292

package, 271
parlay, 126

Pascal's Triangle, 99
Pascal, Blaise, 4
pass line, 60
Peg Solitaire, 266
Penny Ante, 34
perfect matching, 144
permutation, 24
Pigeon Hole Principle, 136
place, 64
point, 81
Poisson Distribution, 115
poker, 49
Poker Dice, 69
Poker Nim, 286
population, 112, 160
positions, 277
Prediction Trick, 151
probability, 9
    event, $P(E)$, 10
probability distribution, 103
Problem of Points, 4, 32
progressively bounded, 294
proper subset, 319
push, 62

qualifying hand, 72
quartile, first, 170
quartile, lower, 170
quartile, second, 170
quartile, third, 170
quartile, upper, 170

raise, 72
random variable, 29
recurrence relation, 247
recursive formula, 247
regression line, 178
regular graph, 257
relative batting average, 226
Robertson, Oscar, 184

roulette, 54
royal flush, 50
Royal Hand, 77
Rubin, Ernest, 159

Sabathia, C.C., 199
sample, 160
sample population, 93
sample space, 5, 319
samples, 112
scanning, 242
scatter plot, 175
set, 319
shooter, 58
short game, 296
significance level, 206
Silver Dollar Game, 289
Simpson's Paradox, 162
Simpson, Edward, 163
sink, 292
slugging percentage, 189
Small Arithmetic Trick, 153
Smith, John, 159
sports stats, 159
Sprague-Grundy function, 293
spread, 31
St. Petersburg Paradox, 33
standard deck, 9
standard deviation, 190
standard deviation, $\sigma(X)$, 43
standard normal distribution, 105
statistics, 159
stemplot, 169
Stockton, John, 185
straight flush, 50
streaks, 120
subtraction game, 278
subtraction set, 279
Sudoku, 242
sum of squares for errors, 178

surrender, 79

terminal position, 277
test value, 206
time plot, 175
Tower of Hanoi, 245
tree diagram, 35
Two-Deck Matching Game, 147
Type I error, 207
Type II error, 207

uniform distribution, 13
uniform probability distribution,
    13
unimodal, 169
union, 320

variance, $var(X)$, 42
vertex, 143
vos Savant, Marilyn, 66

Ward, Cam, 199
weakly magic, 239
West, Jerry, 184

Yule, Udny, 163
Yule-Simpson Effect, 163